HOW SCIENCE WORKS

HOW SCIENCE WORKS
Evaluating Evidence in Biology and Medicine

Stephen H. Jenkins

OXFORD

UNIVERSITY PRESS

2004

OXFORD
UNIVERSITY PRESS

Oxford   New York
Auckland   Bangkok   Buenos Aires   Cape Town   Chennai
Dar es Salaam   Delhi   Hong Kong   Istanbul   Karachi   Kolkata
Kuala Lumpur   Madrid   Melbourne   Mexico City   Mumbai   Nairobi
São Paulo   Shanghai   Taipei   Tokyo   Toronto

Published by Oxford University Press, Inc.
198 Madison Avenue, New York, New York 10016

www.oup.com

Oxford is a registered trademark of Oxford University Press

Library of Congress Cataloging-in-Publication Data
Jenkins, Stephen H.
How science works : evaluating evidence in biology and medicine / Stephen H. Jenkins.
p. cm.
ISBN 978-0-19-515894-6; 978-0-19-515895-3 (pbk.)

1. Biology—Methodology. 2. Medicine—Methodology. I. Title.
QH324.J46 2004
570'.28—dc22   2003049407

9 8 7 6 5 4 3

Printed in the United States of America
on acid-free paper

Preface

I am probably a fairly typical science faculty member at a medium-sized public university in a small state without an ocean as one of its borders. I teach
undergraduate classes, I do research with graduate students, and I publish articles in technical journals on moderately esoteric topics. Until recently, I
never imagined that I would want to write a book. But I've long been interested in the details of the scientific process, from its philosophical under-
pinnings to general issues of experimental design to aspects of modeling and
statistical analysis. I should emphasize that my interests in these topics are
those of a practicing scientist, not of an authority on philosophy or statistics.

In all of my teaching, including beginning biology classes for undergradu-
ates, I've always emphasized depth rather than breadth. I like to talk about
the details of a few key studies, including their assumptions, methods, limitations, and implications for future work, rather than summarizing results from
a large number of studies. Several years ago, I was asked to develop a core
course in research design for graduate students in ecology, evolution, and conservation biology at the University of Nevada, Reno. This further kindled
my interest in diverse aspects of the scientific process, especially because researchers in these fields use a wide range of methods to answer questions. I
developed a large bibliography in fields ranging from philosophy to psychology and from traditional to fairly arcane aspects of statistics. I shared this
bibliography with my students, sometimes to their amazement at the obscure
references I had found, sometimes to their consternation that there was so
much to be learned.

Teaching must be one of the most rewarding professions because it involves a lifetime of opportunities to influence people in small ways and large.

But about 3 years ago I began to think that it would be a good challenge to try to introduce to a larger audience some fundamental ideas about how science works. I also became increasingly concerned about the inadequate and even misleading treatment of these ideas in political discourse, in the news media, and even in basic science education. This seemed especially unfortunate because many of the decisions that individuals and societies must make these days depend on better understanding of how science works. No doubt my advancing age contributed to this concern, and an unknown but shrinking amount of time left on Earth (e.g., I won't be around to see the worst effects of global climate change) motivated me to start this book.

As I gathered ideas for a book on the scientific process for a general audience, I remembered an earlier attempt to promote scientific literacy that got a lot of attention when it was published in 1991: *Science Matters: Achieving Scientific Literacy* by R. M. Hazen and J. Trefil. When I scanned the book, I quickly discovered that it took a very traditional and unsatisfying approach to scientific literacy. The style of *Science Matters* is exemplified by the list on the inside cover entitled "What You Need to Know and Where to Find It." For example, you can turn to page 28 to learn about absolute zero or page 52 to learn about x-rays. If you are curious about x-rays, you get a definition, a description of how an x-ray machine works, and three sentences about how x-rays are useful in medicine and crystallography. Hazen and Trefil present the great ideas of science mostly as a collection of facts that they believe every educated person should know. Many scientists objected to *Science Matters* because Hazen and Trefil, in their emphasis on the end products of science, devoted scant attention to the process of doing science (Culotta 1991; Pool 1991). Nevertheless, the impact of their approach is reflected in the science standards adopted by boards of education in states such as California (Bruton and Fong 2000). These standards typically comprise long lists of facts and concepts followed by briefer lists of topics for "investigation and experimentation" at each grade level.

The science journalist Boyce Rensberger (2000) argued for a different approach, based partly on fascinating results of a recent survey of scientific literacy in the United States and 13 other developed countries (National Science Board 1998). Contrary to popular opinion, "American adults understand basic scientific facts at least as well as those in most other developed countries" (Rensberger 2000:61). The most interesting result of the survey was that knowledge of specific scientific facts among American adults is substantially greater than knowledge of basic elements of the scientific process such as experimentation and hypothesis testing. Rensberger concluded that "the weakness in the public's understanding of science lies in an area not often addressed in interactions between scientists and journalists—the nature of evidence." He recommended that scientists explain their methods when describing research results to journalists so that journalists can include discussions of the evidence for new findings, and he suggested that this might help the 70% of adults in America interested in science (National Science

Board 1998) to distinguish pseudoscience from science more effectively. This encouraged me to think that a new book on scientific literacy emphasizing the process of science and the evaluation of different kinds of evidence might be worthwhile.

I completed most of this book while on sabbatical leave from the University of Nevada, Reno, and I thank the Board of Regents of the University and Community College System of Nevada for granting me this leave. I appreciate the support of my department chair, Lee Weber; dean, Robert Mead; and provost, John Frederick. Thomas Nickles provided stimulating discussion and key references during the early stages of the project.

I am grateful to many people who provided valuable information or helpful reviews of individual chapters: Matthew Jenkins (Chapters 1, 2, 4, and 6); John Basey (2 and 9); Christie Howard (2 and 6); John Worrall (2); Amy Barber, I. Lehr Brisbin, Allen Gardner, and Guy Hoelzer (3); Andrew Blaustein, Wally Jenkins, Pieter Johnson, Joseph Kiesecker, and Eric Simandle (4); Nicola Clayton, Grant Mastick, Vladimir Pravosudov, and Stephen Vander Wall (5); Paul Sherman (6 and 7); Colin Begg and Paul Lichtenstein (6); Joseph Veech (8 and 9); Michael Klag (8); and Pim Martens, Robert Mead, and Sarah Randolph (9). My son Matthew Jenkins and my colleague Christie Howard provided especially valuable suggestions for making my writing more accessible to readers without scientific backgrounds. The detailed comments of Lehr Brisbin, Grant Mastick, Steve Vander Wall, Paul Sherman, and Joe Veech were particularly helpful in improving the organization and focus of Chapters 3, 5, 7, 8, and 9. I also appreciate the comments of Marc Mangel and several anonymous reviewers of my proposal for the book and of an early version of Chapter 6.

Julie Ellsworth and Thomas Nickles read the entire manuscript and provided many specific suggestions, as well as general reactions to the book as a whole. Their enthusiastic response to my writing was heartening.

I thank Grant Hokit for providing the photographs used in Figure 4.1; Kluwer Academic Publishers and Lisa Gannett for permission to use two quotations from "What's in a Cause? The Pragmatic Dimensions of Genetic Explanations" by L. Gannett, in *Biology & Philosophy*, vol. 14, pp. 349–374, copyright 1999; Elsevier Science for permission to use two quotations from "Climate Change and Future Populations at Risk of Malaria" by P. Martens et al., in *Global Environmental Change*, vol. 9, pp. S89–S107, copyright 1999; and the National Association of Biology Teachers for permission to use a quotation from "Vulnerability—The Strength of Science" by Garrett Hardin, in *American Biology Teacher*, vol. 38, pp. 465 and 483, copyright 1976.

I thank my editor at Oxford University Press, Kirk Jensen, for his early encouragement and continuing enthusiasm for this project and for his valuable editorial suggestions.

My father, Dr. Ward S. Jenkins, has spent a lifetime exploring ideas, and I appreciate the example he set. My mother-in-law, Ann O. Pleiss, inquired about my progress weekly, which provided important momentum.

Most especially I thank my wife, Katherine Pleiss Jenkins, for her support and encouragement in many different ways. She was the first reader of each chapter of the book and a gentle but thorough critic who greatly improved the clarity of my writing.

I dedicate this book to the memory of my mother, Elizabeth Howell Jenkins, artist and environmentalist.

Reno, Nevada

Contents

HOW SCIENCE WORKS

Chapter 1

Introduction

Since you are reading this, you probably have some interest in knowing more about how science works. Perhaps you have been intrigued by news reports about science but dissatisfied with their superficiality. You may be a young person wondering about a possible career in science. You may even be a practicing scientist curious about how I will present a topic that is already quite familiar to you.

I hope to stimulate a diverse group of readers, but my main goal is to tell some stories about science that are richer in detail than most science reports in the popular press. Thus I am writing for people who may have great interest in science but little or no technical training and who get most of their information about science from the news media. If you are a beginning biology student, I hope this book whets your appetite for more detailed study of the principles that underlie these stories. If you are a teacher, I hope you find some examples that help you engage your students in thinking more deeply about science. If you are doing research in one of the areas I discuss, you will undoubtedly think of important details I should have included or different points I should have emphasized, but I hope you conclude that my translation of your story for a nontechnical reader is accurate and fair. Why am I writing a book primarily for readers who learn about science from reading or listening to the news, while also hoping that those who use textbooks, reference materials, or in-depth research will find valuable ideas in these pages?

Science stories are regular features of the daily news, although usually not as prominent as stories about war and peace, politics, economics, and especially sports and entertainment. In its coverage of science, my local newspaper, the *Reno Gazette-Journal*, is probably fairly typical of newspapers in all

3

but the largest cities in the United States. I haphazardly selected 13 issues, published between 2 August and 2 September 2002 to examine its coverage of science.[1] After excluding stories that simply reported new cases of West Nile virus, a series of stories about a cancer cluster in a small town near Reno, and the weekly health section, I found 15 science stories, for an average of about one per day. Most of these stories were about human health and nutrition, with headlines such as "Study: Gingko Fails to Give Your Memory a Boost," "New Research Suggests Less Genetic Risk for Breast Cancer," "Experts: Cloned Animals Might Be Safe to Eat," and "Putting Caffeine on Skin Lowers Risk of Cancer in Lab Mice." Others were about geology ("Earth Getting Fatter around the Equator"), global climate change ("Severe Weather Not New, Will Happen Again, Scientists Say"), and evolution ("Researchers: Chimps May Have Survived AIDS Epidemic"). The stories ranged in length from about 100 to 900 words.

Most of these articles were based on newly published papers in scientific journals that caught the attention of science journalists or were promoted by the researchers or their institutions because they were thought to be of some general interest. Another major source of science stories in the popular press is reports released by government agencies or commissions, as in the report about cloned animals being safe to eat. Such stories typically describe some new discovery ("Putting Caffeine on Skin Lowers Risk of Cancer in Lab Mice") or a study that overturns prevailing wisdom ("New Research Suggests Less Genetic Risk for Breast Cancer"). The fundamental problem with almost all of these stories is that they emphasize the conclusions of researchers but give scant attention to the methods used to reach these conclusions. There may be a brief explanation of a key piece of evidence but rarely any discussion of the assumptions made in interpreting this evidence (Rensberger 2000). The reason for this weakness is simply lack of space or time to develop the details of the stories. Science often gets only cursory attention in the media because it competes for attention with many other topics.

There are several unfortunate consequences of the approach to science often taken in the news media. First, it reinforces the belief that science is a unique activity and scientists are fundamentally different from the rest of humanity. This belief can be called the cult of the expert. It is rooted in the assumption that a great deal of technical training is necessary to become a scientist, and therefore scientists are the only ones who can truly understand what other scientists do. Based on the cult of the expert, the role of the news media is to transmit pronouncements of scientists to the general public. Neither reporters nor consumers of news are responsible for evaluating these pronouncements. Instead, when different scientists make contrasting pronouncements, the issue is which scientist has the strongest credentials, and therefore should be considered most credible, or which scientist is the maverick challenging an orthodox view, and therefore should be favored because of his or her status as an underdog.

The cult of the expert has a second negative consequence: news about science can be confusing when successive stories about a topic report different conclusions. For example, a headline one month might say "Red Wine Protects the Heart," while several months later we might read "New Evidence Shows Red Wine Bad for Health." These stories are probably reporting different types of studies of different groups of people done by different researchers. Which conclusion should we believe? The most recent, simply because new research always trumps older research? The study done by a member of the National Academy of Sciences because of the reputation of that senior scientist? On a more practical level, should we drink more red wine or less?

A third unfortunate consequence of science reporting is that readers miss the fun and excitement that were part of the discovery process when stories focus on the end results. To be sure, there are many popular books about science and some in-depth stories in magazines and newspapers that emphasize the quest for a solution to a problem, but most standard news stories are too short to say much about the scientific process. If they mention the excitement of discovery, it's usually in the context of an interview with one of the investigators rather than an explanation of the process of gathering and evaluating evidence. This may enable readers to appreciate the joy of discovery, but it doesn't encourage them to become actively engaged.

Finally, typical news stories about science don't prepare readers to think more deeply about scientific issues. These stories provide lots of information but little education, although they could provide both if they focused more on the scientific process leading to new results. By taking some of the mystery out of science, this would benefit not only individuals who read the news regularly but also society in general. For example, it might promote more rational discussion of social and political decisions that relate to science.

One of my primary motivations for writing this book is to present a more accurate image of science than you would get if you relied exclusively on the news media to keep up with scientific progress. In particular, I will use several contemporary stories in biology and medicine to try to deflate the cult of the expert by describing some of the nitty-gritty details of the thought processes and research methods used by the key players. Modern science is undoubtedly complex, but I believe that many aspects of this complexity are accessible without extensive technical training. Only you can judge whether this belief is justified.

News about science is often confusing because different kinds of evidence point in different directions. It's not yet clear, for example, whether the net effects of a daily glass of red wine are beneficial or detrimental. But this uncertainty is an inherent result of the fact that scientific knowledge grows by fits and starts in unpredictable directions, more like a rambling house with rooms added as the need arises than like an edifice whose final form can be visualized by the original architect. The essence of science is not some nuggets of information about the natural world but rather an ongoing process for gradually learning how the world works, with occasional breakthroughs in the form of major discoveries. At any given time, the understanding of a phenomenon is likely to be incomplete, with conflicting explanations and evidence. Scientists have learned to tolerate such uncertainty and even relish the challenges it offers. Nonscientists will have a better appreciation of the strengths and limitations of science if they can adopt similar attitudes about unresolved scientific issues. The most serious challenge for both scientists and nonscientists is how to make practical decisions in light of ambiguous scientific evidence. This challenge applies to both personal choices about health and nutrition and choices society must make about environmental regulations and other public policies. I will argue that ignoring the scientific uncertainties is not an effective strategy for making these kinds of decisions.

Perhaps my most important goal is simply to show you the pleasure that can be had from thinking rigorously and critically about how scientists try to solve problems. I've selected some stories that I think will capture your attention because they are about fascinating natural phenomena, such as dwindling populations of frogs or the prodigious memory abilities of food-storing birds, or about issues that may be of intense personal interest, such as the causes of cancer. But my underlying goal is to draw you into each story because of the topic itself and then have you discover that the real excitement is in the various approaches of researchers in answering key questions. I hope to show you "why science should warm our hearts," in the words of the philosopher Colin Tudge (2002).

I've scolded the news media for giving an incomplete and inaccurate picture of science. However, I see this book as a supplement to daily news about science, not an alternative. Many science journalists do an excellent job, especially when they have an opportunity to write longer stories about topics that they have investigated in depth. Even the brief reports of scientific progress that are staples of the daily news are an important source of information for general readers. But one goal of this book is to help you read future stories about science in the popular press with more understanding and insight. Even if the author of a story doesn't tell you much about the assumptions of a study or the methods used to test an idea or hypothesis, you may be able to make some educated guesses about these things and interpret the story with an appropriate mixture of enthusiasm and skepticism.

HOW SCIENCE WORKS

One aspect of modern science that contributes to the cult of the expert is the assumption that science depends on complex technology, so it is hopeless for someone without technical training to try to understand the details of the scientific process. It is certainly true that scientists working today have amazing pieces of equipment that make it possible to do things that the previous generation could only dream about and the one before couldn't even imagine. For example, the Human Genome Project, in which the entire genetic code of human beings was determined, would not have been possible without the development of automated sequencing machines for reading the information in long strands of DNA. The human genetic code consists of about 30,000 genes distributed among 24 different chromosomes. The letters of this code are the four nucleotide bases of DNA, and each gene consists of a unique sequence of these bases. In humans, the total number of nucleotide bases on the 24 chromosomes is about 3 billion. Two groups published first drafts of the human genetic code in February 2001: a private company called Celera Genomics and a public consortium of scientists from various universities and government agencies. The private company used 300 DNA sequencers, which cost $300,000 each, plus powerful supercomputers to analyze the data from the sequencers to produce their draft of the human genome in fewer than 3 years (Pennisi 2001; Lorentz et al. 2002).[2]

This example could be multiplied many times, but there are still opportunities to make discoveries the old-fashioned way, by using simple observations. To be sure, this traditional approach depends on hard work, perseverance, detailed knowledge to provide a context for interpreting new observations, and sometimes good luck. For example, Philip Gingerich, Hans Thewissen, and others found a series of fossils in Pakistan and Egypt during the last 30 years that clearly established how whales evolved from even-toed ungulates, the group of mammals that includes cows, sheep, hippos, and related species (Thewissen 1998; Sutera 2000; Wong 2002). Their research involved field-work under very challenging conditions, both environmental and political; painstaking preparation of the fossils; then visual comparison of the various specimens. The sequence of intermediate forms between terrestrial mammals adapted for running and marine mammals with no external limbs, nostrils on the tops of their skulls, and other adaptations for living in water is a truly amazing illustration of an important evolutionary transformation, yet demonstration of this transformation was primarily a low-tech effort.[3]

These apparently very different stories about deducing the genetic code of humans and describing the evolutionary transformation from ungulates to whales have some common features despite their different reliance on complex technology. These common features are mainly ways of thinking about problems, that is, mental tools rather than technological tools. These mental tools are fundamental components of the scientific process, which make science an especially productive way of solving problems. I believe that this aspect of how science works is accessible to anyone willing to exercise his or her brain, regardless of technical background. Therefore, I use examples that illustrate some of the basic analytical methods that underlie all areas of science, rather than examples that show the contribution of gee-whiz technology.

A bit more consideration of the Human Genome Project may help clarify this point. Although decoding enormous quantities of DNA required the development of automated machines and computers capable of analyzing large amounts of data, it also depended on thorough understanding of the structure of DNA and clever experiments to tease apart how DNA molecules are synthesized. The design of these experiments was no different in principle than the design of any experiments in biology and medicine, even ones in which results could be obtained by simple observation of experimental subjects, such as human volunteers in medical experiments. Designing critical experiments to discriminate clearly among alternative hypotheses is essentially the same process whether the hypotheses are about effects of vitamin C on colds (Chapter 2) or about the structure of molecules such as DNA, although experiments on the latter may require highly specialized equipment for obtaining and analyzing results. By telling some scientific stories that don't involve complex technology, I hope to give you some tools for understanding the scientific process that can be applied much more broadly to additional examples, including those in which technology plays a larger role.

PLAN OF THE BOOK

→ REBIRTh

Scientists use two fundamental approaches for answering questions about nature, including human nature: nonexperimental methods involving pure observation and measurement, and experimental methods involving manipulation of natural processes. The key words in this sentence are "pure" and "manipulation" because observation and measurement are critical elements of both experimental and nonexperimental studies. In both types of studies, a comparative framework is usually important for interpreting results. In medical experiments, for example, responses to a new treatment (experimental manipulation) in one group of people might be compared to responses in a control group that did not receive the treatment. In nonexperimental studies, health might be compared in two groups of people with different habits, such as smokers and nonsmokers.

Chapter 2 uses several studies of the health effects of vitamin C and similar compounds to illustrate both nonexperimental and experimental approaches in medical research. Experimental studies in medicine are called randomized, double-blind trials and are often considered the "gold standard" in such research. I introduce this approach by discussing two experiments to test the effects of large doses of vitamin C on the common cold. These examples illustrate some of the basic decisions that must be made in designing any experiment, such as what to use as a control treatment for comparison with the experimental treatment and how to measure responses to the treatments.

Studies of the effects of vitamin C on the common cold demonstrate some of the pitfalls of designing effective experiments. Although the hypotheses being tested in these experiments were straightforward, the procedures were relatively simple, and the analyses of results were uncomplicated, certain aspects of the experiments contributed to uncertain conclusions. I use these studies to illustrate experiments in medicine because their flaws are as informative as their strengths.

In Chapter 2 I also introduce purely observational studies of the long-term effects of vitamin C and similar compounds on aging. One of the key studies asked whether elderly people in Basel, Switzerland, with high levels of vitamin C in the blood had better memory abilities than people with lower levels of vitamin C. In this example, different levels of vitamin C in the blood reflected dietary differences among subjects over long periods of time and possibly genetic differences affecting the metabolism of vitamin C. This and related examples illustrate the ambiguities that arise in interpreting results of

COMPETITION FOR OUTCOMES

nonexperimental studies, which are often more problematic than the pitfalls of interpreting experimental results. However, these examples also show that some kinds of questions don't lend themselves to an experimental approach. I revisit this theme in Chapter 8, which compares experimental and non-experimental studies of the effects of caffeine on blood pressure. Short-term effects were studied with some well-designed experiments, but understanding the consequences of a lifetime of coffee use required a purely observational approach.

I compare experimental and observational methods in several other chapters also. Chapter 3 describes the special challenges and rewards of using experiments to study animal behavior, in particular the ability of dogs to identify individual human beings by smell. Chapter 4 shows how observations in natural environments, laboratory experiments, and field experiments can be integrated to answer ecological questions. Chapter 5 returns to a topic in animal behavior, the spatial memory abilities of food-storing animals. These abilities were tested in a set of clever laboratory experiments with nutcrackers, which complemented observations of the behavior of the birds in nature. In addition, comparative studies of the brains of various species provided insight about the neurological basis of spatial memory. These comparative studies were purely observational, like studies of the long-term effects of caffeine on blood pressure, but the context of the comparisons was much different. In the caffeine studies, researchers were trying to understand differences in health among people who differed in coffee use and other habits; in the neurological studies related to spatial memory, researchers were trying to understand differences among species that have existed for thousands of generations. Nevertheless, the nonexperimental nature of both types of research produces similar challenges in drawing definitive conclusions.

In addition to illustrating how experiments can be designed to test hypotheses in animal behavior, Chapter 3 has two other themes. First, it compares the use of evidence to answer scientific and legal questions. I return to this topic in Chapter 10. Second, Chapter 3 introduces a quantitative approach to evaluating the strength of evidence for or against an hypothesis. For example, we might hypothesize that a particular person committed a crime. Some evidence might include identification of the suspect by a trained police dog in a lineup or DNA of the suspect that matches DNA extracted from blood found at the crime scene. Under some circumstances, the strength of such evidence can be analyzed precisely enough to come up with a numerical estimate of the likelihood of guilt or innocence. But the results of these calculations can also be surprising. Chapter 3 describes the assumptions of this approach and discusses its merits and limitations.

Chapter 4 tells two stories about frogs, one about trying to find the cause of high frequencies of leg deformities and one about trying to understand widespread population declines. I use the word "story" deliberately to emphasize how understanding these two problems developed through a sequence of observations and experiments, with results along the way leading to new hypotheses that were tested in further studies. One of my primary goals in

Chapter 4 is to show how the <u>integration of results</u> from naturalistic <u>observations, laboratory experiments,</u> and field experiments can be a powerful <u>approach to rapid progress in biology.</u> By contrast, reliance on a single method such as controlled laboratory experiments can lead to dead ends. In addition to illustrating the complementary strengths of observational and experimental methods in ecology, the examples in Chapter 4 are important case studies in conservation biology.

Chapter 5 continues exploring the interplay between experimental and observational approaches, but it has a more important objective. Several authors have pointed out the benefits of developing and testing competing hypotheses for a phenomenon. For example, some species of animals store large amounts of food in widely dispersed locations in preparation for a season of food scarcity. How do animals find their stored food weeks or months later? Several possible mechanisms can be imagined, ranging from using smell to detect stored items (which may be invisible because they are buried) to remembering specific locations of the stored food. Chapter 5 discusses a set of simple but ingenious experiments to discriminate among these and other hypotheses. I use this concrete example to illuminate some fundamental philosophical principles about constructing and testing alternative hypotheses and about the roles of positive and negative evidence in science.

Chapters 6 and 7 don't focus as closely on observational and experimental methods as the other chapters, but they develop some ideas about causation that are important in interpreting studies in biology. I return to a medical example in Chapter 6: how the risk of getting cancer is influenced by genetic and environmental factors. This chapter was motivated by contrasting news accounts that appeared in July 2000 of a large Scandinavian study of cancer incidence in twins. One account highlighted the importance of genetic contributions to cancer risk; the other emphasized the importance of environmental factors. I was surprised that these stories, which appeared in major U.S. newspapers, could present such different interpretations of the same scientific study, so I read the original report in the *New England Journal of Medicine.* This led to the main theme of <u>Chapter 6: the complex and diverse ways in which causal factors can interact</u> to <u>influence biological processes.</u> I won't give away the conclusion about the roles of genetic and environmental factors in cancer here because the fascinating details are important for fully appreciating this conclusion.

This example was an important stimulus for me to write this book. It made me think about how science is presented in the press and whether it might be useful to elaborate on several recent news stories to show how science really works. I found that I could use different stories to illustrate various fundamental points about scientific methods, and eventually I had an outline for the book.

Chapter 7 continues to explore the complexities of causation, using <u>different hypotheses about the causes of aging to illustrate a central theme</u>—that biological phenomena have complementary causes at multiple levels: biochemical, physiological, genetic, developmental, environmental, and evolutionary. This theme contrasts with that of Chapter 5, which shows the power of testing alternative, competing hypotheses to answer questions in biology. But the hypotheses considered in Chapter 5 are all attempts to explain the behavioral *mechanisms* by which birds and rodents find food that they store; that is, they are all explanations at the same level. In Chapter 7, I discuss mechanistic hypotheses about the biochemical and physiological causes of aging, as well as hypotheses about environmental factors that contribute to aging, genetic and developmental aspects of aging, and evolutionary reasons for aging. These kinds of hypotheses at different levels are not alternative explanations in the sense that if one is correct, the others must be false. Instead, all of these levels of causation must be considered for a full understanding of aging and of biological phenomena in general. Furthermore, progress in understanding at one level, such as mechanisms of aging, can stimulate progress in understanding at other levels, such as evolutionary reasons for aging.

Chapter 8 uses some studies of the effects of coffee on blood pressure and on cancer that nicely illustrate the tradeoffs between experimental and purely observational research in medicine. This chapter also introduces the key role of replication in scientific research and describes a method that is commonly used to determine whether multiple studies of the same question give consistent answers. This quantitative approach to combining results from different studies is called *meta-analysis*. Many news stories about medical and nutritional research actually report the results of such statistical summaries of experimental or observational studies of a topic, without identifying them as meta-analyses or explaining how the authors of the summaries reached their conclusions. Because of the widespread use of meta-analysis in medicine and nutrition, as well as in education, sociology, and other branches of the social sciences, I believe it is useful to understand the basic elements of this method.

Chapter 9 explores another basic conceptual tool of science, the use of quantitative models to help crystallize ideas and hypotheses. The primary example in this chapter is the possible effects of global climate change on the occurrence of malaria in various parts of the world. Many people are intensely interested in predicting the likely consequences of global climate change during the next 50 to 100 years. These consequences include many changes in natural environments, as well as potential effects on human health. Because the global climate system is extremely complex, quantitative computer models have played a key role in developing predictions about the extent and possible consequences of climate change. This is a topic for which it is tempting to defer to the experts who use these models. However, the predictions of the models have significant implications both for personal lifestyles and for national and international policies about energy use and other environmental issues. Furthermore, these predictions have been hotly contested by some scientists, which has promoted confusion about global climate change among politicians and members of the general public. Therefore, it's important to have a basic understanding of these models so that those without technical training can make intelligent judgments about the associated controversies.

There are various approaches to modeling, and a major goal of Chapter 9 is to compare two very different kinds of models of the future distribution of malaria. These models have contrasting strengths and limitations. By giving you an opportunity to compare two such models, I hope to demystify the process of modeling by showing how the predictions of any model are linked to the assumptions used in making it. My intention is not to contribute to your possible distrust of abstract models in general but to help you develop some tools for discerning the strongest points of particular models, as well as their most significant weaknesses. The process of comparing two different models should be an effective way to do this.

In Chapter 10 I elaborate on several threads that are initiated in the questions considered in earlier chapters. Most of the questions do not have final answers despite good and productive research directed toward them, partly because of their nature: human health (Chapters 2, 6, 7, and 8), animal behavior (Chapters 3 and 5), and global ecology (Chapters 4 and 9) are inherently complex because of the great diversity of factors that can influence them and the variability in responses to these factors that exists in humans and other species. The lack of definitive answers to questions discussed in this book also reflects the fact that science is an ongoing process in which the most important sign of progress is often that results of an experiment or observational study lead to a new set of questions. This is part of what makes science exciting and rewarding for scientists, but it entails an important dilemma: how do we make the best practical and even ethical decisions based on incomplete scientific knowledge? Science impinges on many of the decisions that individuals and society in general must make. But there is often a fundamental tension between the tentativeness of scientific conclusions and the necessary finality of some of our practical and ethical decisions.

Science is only one way of gaining understanding of the world, albeit an especially powerful way. How does the scientific approach compare to other approaches, such as art and religion? What are the strengths and limitations of these different approaches? What do they have in common? On a more practical level, how does the use of evidence by scientists and lawyers differ? I take up these kinds of questions in Chapter 10 to explore the scope of science in the larger context of the multiple ways in which we, as humans, deal with the world.

Finally, in Chapter 10, I discuss the interplay between two key traits of scientists, curiosity and skepticism. In many ways these are contrasting human traits: consider the pure boundless curiosity of a young child and the unmitigated distrust of an old curmudgeon, for example. But the key to success as a scientist is often maintaining a delicate balance between these two traits. In this book I hope to encourage both the curiosity and skepticism of readers who are not trained as scientists. I believe this will help you read and interpret science news with greater understanding and pleasure and make more satisfying personal decisions about issues affected by developments in science based on knowing more about how science works.

RECOMMENDED RESOURCES

In addition to the references for specific information and ideas in each chapter that are collected at the end of the book, I provide a few recommended readings and online resources at the end of individual chapters. These include articles and books that I found especially valuable and current web sites that deal with critical thinking and the scientific process. Some of the books published in the 1990s and earlier are little gems, packed with insight, that deserve more attention than they've received; I promote them unabashedly.

Best, J. 2001. *Damned lies and statistics: Untangling numbers from the media, politicians, and activists.* University of California Press, Berkeley. Best illustrates the importance of a critical approach to statistics reported in the news media.

Coggon, D., G. Rose, and D. J. P. Barker. 1997. *Epidemiology for the uninitiated,* 4th ed. http://bmj.com/epidem/epid.html (accessed April 3, 2003). This site, sponsored by the *British Medical Journal,* is a good source of basic descriptions of different ways of designing research and analyzing results of research in medicine. The general principles apply broadly to all kinds of research in biology.

Herreid, C. F., and N. A. Schiller. 2003. National Center for Case Study Teaching in Science. http://ublib.buffalo.edu/libraries/projects/cases (accessed November 30, 2002). Herreid and Schiller advocate a case study approach to teaching science and provide many examples at this web site.

Paulos, J. A. 1995. *A mathematician reads the newspaper.* Basic Books, New York. This is a delightful little book about a great variety of ways that mathematical ideas appear in the news media.

Rensberger, B. 2000. The nature of evidence. *Science* 289:61. Rensberger's article was an important stimulus for writing this book.

Chapter 2

Do Vitamin C and Other Antioxidants Benefit Health?
Using Observational and Experimental Studies to Test Medical Hypotheses

Linus Pauling was a creative and prolific chemist who was awarded two Nobel Prizes: the Chemistry Prize in 1954 for fundamental work on the structure of molecules and the Peace Prize in 1963 for articulating the dangers of nuclear proliferation. He might have become the only person to receive three Nobels if he had beaten Watson and Crick in deducing the structure of DNA. Late in his long life, Pauling became a proponent of the multiple health benefits of large daily doses of vitamin C. I remember hearing Pauling lecture about vitamin C to a large and enthusiastic audience at the University of Nevada, Reno, in 1985, when he was 84 years old. He was a persuasive advocate who used charm and humor, as well as an arsenal of data and anecdotes, to deflect criticism. He summarized his ideas in a book called *Vitamin C and the Common Cold*, first published in 1970. Many books of nutritional advice for the general public have been published, but his has had a staying power matched by few others. One reason may be that Pauling's stature as a scientist gave his ideas automatic credibility. Pauling's expertise in biochemistry makes it tempting to accept his views about vitamin C uncritically, but it is not necessarily wise to do so. His ideas have stimulated many attempts to test various hypotheses about the beneficial effects of vitamin C, and this research illustrates the strengths and limitations of purely observational, as well as experimental, approaches to medical questions. By looking closely at the evidence yourself, I hope you will see the value, as well as the pleasure, that comes from carefully dissecting a scientific problem.

Inadequate intake of vitamin C leads to a disease called scurvy, whose symptoms include dry skin; poor healing of wounds; bleeding from the skin, joints, and gums; and unusual fatigue. Scurvy was especially prevalent among sailors on long sea voyages. Although vitamin C deficiency as the specific cause of scurvy was not discovered until 1911, the British navy learned in the late 1700s that it could easily be prevented by eating fresh fruits and vegetables. Scurvy is uncommon in developed countries today except among alcoholics and individuals who are mentally ill or socially isolated.

It only takes about 10 milligrams (mg) of vitamin C per day to prevent scurvy. The daily intake recommended by the Food and Nutrition Board of the U.S. National Academy of Sciences is 90 mg for males and 75 mg for females (Food and Nutrition Board 2000). These values are sufficient to prevent scurvy, even if a person did not ingest any vitamin C for a month, and may have other health benefits not specifically documented by the Food and Nutrition Board, but they are well below levels that are widely believed to be beneficial. For example, my personal physician recommends a vitamin C supplement of 500 mg/day to slow aging and protect the heart and vascular system, and a recent survey showed that two-thirds of people in the United States who sought medical care for colds believed that vitamin C alleviated their symptoms (Braun et al. 2000). What evidence exists for these and other purported health benefits of large doses of vitamin C?

Many researchers have reported such evidence from various types of studies, so vitamin C is a good example to introduce the array of methods that can be used to learn how health and disease are influenced by diet, as well as by different medical treatments. A critical evaluation of the present state of knowledge should provide a foundation for sensible interpretation of future studies that are reported in the press, especially if we focus on the strengths and limitations of various methods that can be used to answer questions about diet and health because these fundamental methods will be used in future studies as well.

ANTIOXIDANTS AND AGING

Although Linus Pauling did not discuss the possibility that vitamin C might protect against mental deterioration associated with aging in *Vitamin C and the Common Cold*, this possibility has become one of the most popular reasons for taking large daily doses of vitamin C and other antioxidants such as vitamin E and beta-carotene (β-carotene). The rationale is rooted in the chemical reactions of these compounds with *reactive oxygen species*. Cells of living organisms contain energy factories called mitochondria, which carry out the final steps in the conversion of energy to a form that can be used to do the work of the cells. These steps require oxygen, which we obtain by breathing. The process is quite efficient, but small amounts of oxygen are converted to byproducts such as hydrogen peroxide, ozone, nitric oxide, superoxide radicals, and hydroxyl radicals. These substances are called reactive oxygen species because they are derived from oxygen and react readily with essential cell constituents such as DNA, proteins, and lipids. By altering these constituents, reactive oxygen species may cause mutations that lead to unrestrained cell division and growth, that is, cancer, or they may cause cell death and thus contribute to disease (Evans and Halliwell 2001).

Cells have various natural mechanisms to protect them from these effects, but damage from reactive oxygen species may accumulate over time, contributing to the gradual deterioration called aging (see Chapter 7 for a discussion of this and other hypotheses about aging). Vitamins C and E and β-carotene are called antioxidants because they interact with reactive oxygen species, thus reducing the potential for damaging reactions of these substances with DNA, proteins, and lipids in cells. Because the brain has a high rate of oxygen use and abundant lipids in the cell membranes that form connections between cells, brain cells may be particularly vulnerable to the damaging effects of reactive oxygen species. Therefore, it is quite reasonable to hypothesize that increased amounts of antioxidants in the diet might protect against the mental deterioration of aging.

One common way of designing medical and nutritional research is called a *prospective design*. This is illustrated by a long-term study of residents of Basel, Switzerland, begun in 1960. Suppose we hypothesize that antioxidants protect against memory loss with aging. One way to test this hypothesis would be to initiate a study of young to middle-aged individuals who differ in their use of antioxidants or their blood levels of antioxidants *before* decreased memory ability associated with aging is expected to occur. Ideally, this group would represent a random sample of a population of interest and few subjects would drop out of the study before its conclusion, which might be many years later. This is a prospective study because we are looking forward in time to make a prediction about a future outcome (increased memory loss) that may result from a current condition (low intake of antioxidant compounds). A group of Swiss researchers led by W. J. Perrig tested the memory abilities of 442 participants in the Basel study in 1993, when they were between 65 and 94 years old (Perrig et al. 1997). Antioxidant levels had been measured in the blood of these individuals in 1971, independently of any assessment of their memory ability. Therefore, this qualifies as a prospective study.

Researchers at the National Cancer Institute (NCI) did a classic prospective study of the health effects of smoking in the 1960s (Giere 1997), which provides a model for other researchers, such as the Swiss group that studied antioxidants and memory. From more than 400,000 men who volunteered for the smoking study, the researchers selected about 37,000 smokers and an equivalent number of nonsmokers for comparison. The key strength of this large-scale study was that each smoker was paired with a nonsmoker, and members of these pairs were similar in obvious characteristics such as age and ethnic group, as well as a host of other characteristics that might influence health, ranging from religion to the average amount of sleep per night. Not surprisingly, the death rate of smokers after only 3 years was twice that of nonsmokers.

There are several pitfalls of prospective studies compared to the kinds of randomized experiments that have been used to study the effects of vitamin C on the common cold, as described later in this chapter. One of the most significant risks in a prospective study is that individuals being compared may differ in other factors besides those hypothesized to produce an effect. If the frequency of the effect differs between two groups, such as smokers and non-smokers, the reason for this difference may not be the factor the study was designed to test but another factor that also differs between groups, often called a confounding factor. This is why the extensive matching of smokers and nonsmokers was so important in the study by the NCI: it was an attempt to minimize the possibility that differences in death rates could be due to con-founding factors. This matching required a massive effort to recruit 400,000 volunteers to get 74,000 subjects for intensive study. Of course, in principle, it's impossible to measure and control for all conceivable confounding fac-tors. However, it would obviously be both unethical and impractical to study the health effects of smoking experimentally, by randomly assigning individuals to smoke or not smoke for many years and then measuring their survival rates.

The study of antioxidants and aging by Perrig's group differed from the NCI study because the subjects in the former were not divided into two groups but varied along a continuous scale in their blood concentrations of vitamin C, vitamin E, and β-carotene. This variation was evidently not due to differences in the use of vitamin supplements by the subjects because only about 6% stated that they used supplements, and these individuals did not necessarily have high levels of the vitamins circulating in their blood. One of the most interesting results of this study was that blood plasma levels of each of these vitamins in 1971 were strongly correlated with plasma levels in the same individuals in 1993. This might reflect dietary differences among indi-viduals that remained consistent for more than 20 years or genetic differences that affected how vitamins are processed and stored in the body.

Perrig's group used five standard tests of memory performance. One of these was fairly obscure and had no relationship with blood levels of antioxi-dants, but the other four were more useful in testing this relationship. *Im-plicit memory*, or *priming*, was tested by showing subjects a picture containing several familiar objects on a computer screen,[1] then showing the subjects in-dividual pictures of 15 of these familiar objects randomly interspersed with 15 new objects. On average, subjects named familiar objects 17% faster than new ones simply because of the priming effect. *Free recall* was tested by asking the subjects to name as many of the objects in the initial picture as possible after a 20-minute delay. On average, subjects recalled 8.2 objects. *Recognition* was tested after the free-recall assignment by showing a picture containing some old objects from the initial picture and some of the new objects that were used in the priming phase. Subjects were asked to identify the objects that were in the initial picture, and the researchers calculated an index of recognition ability based on the number of correct choices and errors. Finally, *semantic memory* was tested by asking subjects to define 32 words; the average subject got 19.6 definitions correct.

The essential results of this study can be summarized in a set of *correlation coefficients*, which express the relationships between each of the antioxidants (and a few other physiological measurements of the subjects) and each of the memory tests. There are 45 of these correlation coefficients because there were five memory tests done in 1993 and nine physiological measurements, which included levels of vitamin C, vitamin E, and β-carotene, measured in 1971 and 1993, and blood pressure, cholesterol, and ferritin (an iron-containing compound that may reverse the antioxidant effects of vitamin C), measured in 1993.

If we wish to make our own interpretation of the results of the study, rather than simply rely on the authors' conclusion that "ascorbic acid [vitamin C] and β-carotene plasma level are associated with better memory performance" (Perrig et al. 1997:718), we need to consider what can really be learned from correlation analysis, as well as the limitations of this fundamental statistical tool. The correlation coefficient measures the relationship between two variables, such as blood plasma level of β-carotene and ability to recognize objects seen recently, on a scale from −1 to +1. The best way to think about the meaning of this coefficient is to visualize the relationship graphically. If two variables are completely unrelated, then the correlation coefficient equals 0. Figure 2.1A illustrates this pattern (more accurately, lack of pattern). In other words, Figure 2.1A shows what the relationship between plasma level of β-carotene and recognition ability would look like if there were no relationship between these two variables.

By contrast to the random cloud of points shown in Figure 2.1A, which corresponds to a correlation coefficient close to 0, a perfect linear relationship between two variables would have a correlation coefficient of +1 if larger values of one variable corresponded to larger values of the other (Figure 2.1C) or −1 if the reverse were true.

What does the relationship between the plasma level of β-carotene and the recognition ability of the 442 elderly subjects actually look like in comparison to the reference patterns shown in Figure 2.1A and 2.1C? The researchers didn't provide their raw data, but they stated that the correlation coefficient for β-carotene measured in 1993 and recognition ability measured in 1993 was 0.22. Figure 2.1B illustrates a relationship between two variables that are correlated at a level of 0.22.[2]

Does Figure 2.1B make you skeptical of the association between β-carotene and memory ability claimed by Perrig's group? Although I made these graphs to help you think critically about the results of this study, I hesitate to encourage undue skepticism because, as an ecologist, I frequently work with relationships that are just as messy. Chemists, molecular biologists, and even physiologists are used to working with relationships that look like Figure 2.1C, and they regularly scoff at data that look like Figure 2.1B. On the other hand, ecologists, psychologists, and medical scientists have learned how to deal with patterns like Figure 2.1B because they are interested in phenomena

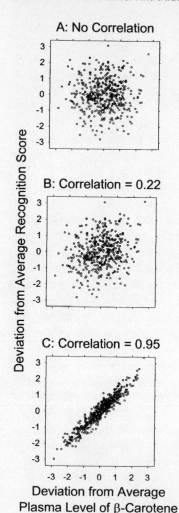

A: No Correlation

B: Correlation = 0.22

C: Correlation = 0.95

Deviation from Average Recognition Score

Deviation from Average Plasma Level of β-Carotene

CORRELATION COEFFICIENT.

Figure 2.1. Hypothetical correlations between plasma levels of β-carotene and scores on a test of the recognition component of memory ability for 442 elderly Swiss studied by Perrig and his colleagues (1997). A shows what the relationship would look like if these variables were uncorrelated (my simulation actually produced a very low correlation coefficient of −0.002), B illustrates a relationship with the correlation coefficient of 0.22 that was in fact obtained by Perrig's group, and C shows a hypothetical relationship with a correlation coefficient close to the maximum value of 1.0. The scales of both axes are in standard deviation units; for example, 1 on the horizontal axis represents 1 standard deviation above the average plasma level of β-carotene in the population, and −1 represents 1 standard deviation below the average (see note 2).

that are influenced by many different factors. If a single response variable such as recognition ability is plotted against a single explanatory variable such as β-carotene, it would be surprising to obtain a relationship as tight as that in Figure 2.1C because of the many other factors that may also influence recognition ability. Much of the scatter in Figure 2.1B is probably due to variation in these other factors: age, sex, education of the subjects, and so on.

In fact, we can go beyond simply visualizing the data and ask the following question: what is the chance of obtaining a correlation coefficient as large as 0.22 if we measure two variables that are truly independent of each other in a group of 442 individuals? This probability can be estimated by randomly generating two sets of unrelated data with 442 items in each set, pairing the items in the two sets arbitrarily and calculating the correlation coefficient between them, repeating this process many times, and counting the number of times that the correlation coefficient is greater than 0.22 in these randomization trials. When I did this numerical experiment, I got no correlation coefficients greater than 0.22 in 1,000 trials, implying that the probability of getting a correlation coefficient as large as 0.22 by chance is less than 1 in 1000. This very low probability enabled Perrig's group to claim a "statistically significant" association between plasma level of β-carotene and the recognition component of memory ability.

Recall that Perrig and his colleagues measured nine physiological variables and five aspects of memory performance. How does the correlation coefficient of 0.22 for recognition ability with plasma level of β-carotene in 1993 compare to the other 44 correlation coefficients among these variables? I purposely picked the relationship between recognition ability and β-carotene to graph in Figure 2.1B because it had the highest correlation. The next highest value was 0.16 for semantic memory ability with plasma level of vitamin C in 1993, and there were 11 correlation coefficients in all that Perrig's group judged to be statistically significant because probabilities of getting these values by chance if the variables were *not* related were less than 5%.[3] There were no significant correlations between the plasma level of vitamin E and the scores on any of the memory tests. Five of the significant correlations involved β-carotene—for levels measured in 1993 with free-recall ability, recognition ability, and semantic memory and for levels measured in 1971 with the latter two aspects of memory. Since the memory tests were only done in 1993, it's noteworthy that 1971 levels of β-carotene were correlated with these aspects of memory measured 22 years later. There were three significant correlations for vitamin C—for levels measured in 1993 with free-recall ability and semantic memory and levels measured in 1971 with semantic memory. The other three significant correlations involved cholesterol with one component of memory ability and ferritin with two components.

These results seem consistent with the authors' conclusion that vitamin C and β-carotene enhance memory performance, but there is one further complication that we need to consider before accepting this interpretation. To set the stage, imagine flipping a coin 10 times. What are the chances that the coin would come up heads every time? This probability can be readily calculated as $0.5^{10} = 0.1\%$. Now imagine that 100 people each flip a coin 10 times. What are the chances that at least one of these people gets 10 heads? This probability turns out to be about 10%.[4] In other words, unlikely events may occur, given enough opportunities. With this in mind, let's reconsider the full set of 45 correlation coefficients between physiological measurements (including blood levels of antioxidants) and components of memory ability reported by Perrig's group. Suppose each of the nine physiological variables was independent of each of the five memory variables. In this hypothetical situation with no real relationships between antioxidant levels and memory ability, what would be the chances of getting correlation coefficients as high as those actually observed? I wrote a small computer program to simulate this thought experiment and found that there was about a 12% probability that the maximum correlation coefficient in a set of 45 would be greater than 0.14

(the three largest correlation coefficients for the actual data of Perrig's group were 0.22, 0.16, and 0.14). In 80% of 1,000 trials with my program, at least one correlation coefficient in a set of 45 was greater than 0.10 (this was the level judged to be statistically significant by Perrig and his colleagues). The results of Perrig's group are still meaningful because they got 11 correlation coefficients greater than 0.10 out of the 45 that they tested, whereas I never got more than six and typically only one or two in my simulation with random data. However, the simulation shows why we should be cautious in interpreting large sets of correlation coefficients, especially when relationships are likely to be influenced by several unmeasured variables (look again at the scatter in the relationship between the plasma level of β-carotene and recognition ability shown in Figure 2.1B). In fact, there is a statistical tool marvelously named the "sequential Bonferroni technique" that was designed to deal with the problem raised in this paragraph (Rice 1989). Applying this technique to the 45 correlation coefficients presented by Perrig's group shows that we can only be confident that two of the correlations represent real relationships between plasma levels of antioxidants and components of memory: the relationship between β-carotene measured in 1993 and recognition memory and the relationship between vitamin C measured in 1993 and semantic memory.[5]

I've expounded at length on correlation analysis only to reach a somewhat dissatisfying conclusion: although we shouldn't embrace the conclusions of Perrig's group wholeheartedly, neither can we categorically dismiss the hypothesis that antioxidants enhance memory performance. My ulterior motive for this extended discussion was to give you some insight about the workings of this widely used statistical technique. But there is a more fundamental problem with the data provided by Perrig and his colleagues that is rooted in the fact that they used a prospective design rather than a controlled experiment. Recall that the Swiss subjects in this study varied in their blood levels of antioxidants because of some unknown combination of factors not determined by the researchers: they may have had genetic differences that influenced their ability to absorb or retain particular vitamins, they undoubtedly had different diets, and so on. How might these other factors influence the results? The analyses discussed so far don't provide a way to answer this question, so Perrig's group used another statistical method called regression, which allowed them to consider multiple explanatory variables simultaneously. They did three regression analyses, using free-recall ability, recognition ability, and semantic memory because each of these components of memory was correlated with one or more antioxidant measures in the initial correlation analysis. I'll illustrate these regression analyses with semantic memory because these results were clearest.

Perrig's group essentially wanted to know if the correlations between scores on the vocabulary test of semantic memory and blood plasma levels of antioxidants could be attributed to other variables that might also be correlated with levels of these antioxidants. Therefore they used educational level, age, and sex as additional explanatory variables together with plasma levels of vitamin C and β-carotene in their regression. If, for example, more educated subjects tended to have higher plasma levels of vitamin C and better semantic memory than less educated ones, the association between vitamin C and semantic memory might simply be an artifact of the relationship between education and these two variables. Not surprisingly, they did discover that subjects with more education did better on the vocabulary test. Younger subjects also performed better, but there was no difference between males and females. However, the regression analysis made it possible to statistically control for the effects of these variables; that is, to ask what the relationship between vitamin C or β-carotene and semantic memory would be if age and level of education were fixed.

Perrig's group found that these relationships were still significant even when controlling for education, age, and sex. Under these conditions, the probability that the correlation between plasma level of vitamin C and semantic memory ability was due solely to chance was 3.4%, compared to less than 0.1% in the initial correlation analysis, which didn't account for education, age, and sex. For β-carotene and semantic memory, the probability that the correlation was due to chance was 3.5% when education, age, and sex were controlled. Perrig's group got similar results for β-carotene and recognition ability but no significant relationships between vitamin C and recognition ability or between any of the antioxidants and free-recall ability.

Regression is more powerful than simple correlation analysis for revealing patterns in data collected in prospective studies because we can consider several variables simultaneously in regression, but regression analysis doesn't overcome the fundamental limitation of these kinds of nonexperimental studies. This limitation is that there may be *unmeasured* factors that account for variation in a response variable (e.g., an aspect of memory ability), as well as variation in a factor that we do measure (e.g., blood level of vitamin C or another antioxidant), producing an artifactual relationship between the factor we are interested in and the response variable.

In the case of the elderly Swiss population studied by Perrig and his colleagues, a causal relationship between antioxidants and memory ability seems more plausible because of the known physiological effects of antioxidants on cells, but it's certainly possible that some other factor, unmeasured by the researchers, was the real reason for differences in memory performance among the subjects. For example, individuals may have had higher levels of vitamin C and β-carotene in their blood because they consistently ate more fruits and vegetables, but some other component of this diet caused differences in performance on memory tests. Or perhaps differences in socioeconomic status or lifestyle caused both differences in dietary intake of antioxidants and differences in memory ability, creating a spurious relationship between plasma levels of antioxidants and performance on memory tests. The possibilities can be multiplied almost endlessly. In considering only education, age, and sex as possible confounding variables, Perrig's group wasn't very thorough in investigating alternative explanations for differences in memory ability among their elderly subjects. Contrast this study with that by the National Cancer

Institute of mortality associated with smoking, in which at least 23 additional explanatory variables besides smoking itself were considered. Although the results of Perrig's group are interesting, they fall far short of being conclusive evidence that antioxidants are beneficial for cognitive performance.

Mohsen Meydani (2001) of Tufts University reviewed studies of the effects of antioxidants on cognitive abilities of older people in an article published in *Nutrition Reviews*. In addition to the study by Perrig's group of elderly people with normal cognitive function, Meydani discussed several studies of individuals with Alzheimer's disease and vascular dementia.[6] Two of these studies used a different design that is even more problematic than the prospective design of the study by Perrig's group, although it is quite common in medical research. This is a *case-control* design in which subjects who already have a particular condition, such as a disease, are compared to a set of control subjects without the condition. In this situation we are trying to deduce the cause of the disease by identifying factors that differ between cases and controls. This may involve simply measuring physiological or other characteristics of the two groups at the time of the study or asking individuals about their past habits (e.g., smoking) or potential exposure to environmental toxins. For example, A. J. Sinclair and four colleagues (1998) studied plasma levels of antioxidants in 25 patients with Alzheimer's disease, 17 patients with vascular dementia, and 41 control subjects without evidence of either disease. The control subjects were similar in age to those with Alzheimer's disease and vascular dementia, but surprisingly the groups were not matched for sex: 36% of the diseased individuals were females, but 59% of the control individuals were females. Sinclair's group found that average plasma levels of vitamin C were about the same in patients with Alzheimer's disease as in controls but were 22% less in those with vascular dementia. Conversely, vitamin E levels were 14% less in patients with Alzheimer's disease than in controls. Sinclair's group concluded, "Subjects with dementia attributed to Alzheimer's disease or to vascular disease have a degree of disturbance in antioxidant balance which may predispose to increased oxidative stress. This may be a potential therapeutic area for antioxidant supplementation" (1998:840).

This study dramatically illustrates some of the fundamental weaknesses of the case-control design compared to prospective designs and experimental studies. Even if the relationships between plasma levels of antioxidants and dementia reported by Sinclair's group are biologically meaningful, there is no way to tell from their data if low vitamin C contributed to vascular dementia, and low vitamin E to Alzheimer's disease, or if these low levels were consequences of the diseases. The cases studied were not a random sample of individuals with Alzheimer's disease and vascular dementia but were patients in a particular medical facility in England, and there is no way of knowing whether or not these individuals are representative of any particular population of people with dementia. In case-control studies, there are ample opportunities for bias in the selection of controls. For example, in the research by Sinclair and his colleagues, the control group had more females than did the group with dementia, suggesting that individuals in the control group were chosen mainly for convenience rather than to match cases and controls as closely as possible. Even without conscious bias, it's extremely difficult to select an appropriate control group in a case-control study because there is a host of potential confounding factors that should be matched between the cases and controls. Such matching is essentially impossible with the small sample size of the study by Sinclair and his colleagues. More important, just as with nonexperimental prospective studies, there is always the possibility that unmeasured or even unimagined variables account for differences in disease between cases and controls.

In his 2001 review, Meydani concluded that antioxidants protect against deterioration of cognitive ability with age and that taking high doses of supplements such as vitamin E might be beneficial to forestall this symptom of aging. I've discussed two of the studies that led to this conclusion, primarily to illustrate the limitations of nonexperimental studies in answering questions about human health. The other research described by Meydani includes experimental studies with animals and even an experimental study of vitamin supplementation for Alzheimer's patients, but it is no more convincing than the two studies that I considered in detail. For example, the authors of the experimental study of Alzheimer's patients had to do major statistical contortions to find a relationship between treatment with large doses of vitamin E and rate of deterioration of the patients (Sano et al. 1997).

Therefore, Meydani's conclusion is questionable despite the plausibility of the proposed mechanism by which antioxidants might help preserve brain function. Nevertheless, my point is not so much to debunk the widely believed notion that antioxidants protect against aging (I myself take 400 units of vitamin E daily, just in case) but rather to set the stage for discussion of a contrasting research strategy that is often considered the gold standard of medical research: randomized, double-blind, experimental trials. For examples of this approach, I'll use tests of the hypothesis that large doses of vitamin C minimize the severity of common colds. After describing two examples, I'll return to the general issue of experimental versus nonexperimental approaches in medicine and cast a more positive light on the kinds of nonexperimental studies that we've considered so far.

DOES VITAMIN C PROTECT AGAINST THE COMMON COLD?

The hypothesis that vitamin C is beneficial in preventing or treating colds has been tested in dozens of experiments dating back to at least 1942. This hypothesis is a natural candidate for experimental testing because effects of supplemental vitamin C should be manifested fairly rapidly in reduced incidence or severity of colds if the hypothesis is true. By contrast, cognitive impairment with age may be a long-term consequence of multiple factors, including use of antioxidant vitamins, that act over decades of life. It's generally not feasible to design rigorous experimental studies on these time scales, so other kinds of evidence have to be used to test hypotheses about long-term effects.

Linus Pauling (1970) discussed a handful of experimental studies that were done before 1970 in his book *Vitamin C and the Common Cold* (see also Pauling 1971). However, the reports of these studies have various problems that make them poor examples for our consideration. The most important problem is that the original reports don't include key information about experimental design, statistical analyses, or numerical results, so they can't be thoroughly evaluated. Therefore, I'll start with an influential study by Thomas Karlowski and five colleagues at the National Institutes of Health (NIH) and reported in the *Journal of the American Medical Association* (*JAMA*) in 1975 (Chalmers 1975; Karlowski et al. 1975). This is a good illustration of the experimental approach, not because it was a flawless study, but for exactly the opposite reason: it had a cardinal flaw, as the authors themselves recognized once the study was underway. In fact, Thomas Chalmers, who was director of the Clinical Center of NIH in the 1970s and one of the authors of the *JAMA* article, reported that he was "more proud of it than almost any other that I have published" (Chalmers 1996:1085), partly because the NIH researchers identified the flaw and were able to account for it in their interpretations of the results.

The general hypothesis that vitamin C helps fight colds actually comprises two specific hypotheses: that taking high doses of vitamin C on a regular basis helps prevent colds and that taking high doses of vitamin C at the first signs of a cold reduces its length and severity. In other words, vitamin C may have a prophylactic effect, by preventing colds, and/or a therapeutic effect, in treating colds. Both of these hypotheses were tested by the NIH group.

Karlowski and his colleagues recruited 311 volunteers for their study from among 2,500 NIH employees. About 600 of these employees indicated willingness to participate, but about half were excluded for various reasons such as health problems that might be exacerbated by taking large supplemental doses of vitamin C, pregnancy, or unwillingness to refrain from taking vitamin supplements outside of the study. The researchers began the study in late summer to capitalize on the fact that frequency of colds in the Washington, D.C., area increases in fall and winter. They planned to continue it for 1 full year, but participants gradually dropped out, so it was ended after 9 months, when the total number of participants fell below 200. A small but significant aspect of the research design was that this stopping rule was decided beforehand. If this had not been the case, the researchers could have been accused of stopping the study when the results were most favorable to their preferred hypothesis.

Standard procedure in studies like these is to compare a group of subjects who receive a treatment with a control group of subjects who do not receive the treatment. In this case, treatment refers to a regimen of vitamin C capsules provided to the subjects with instructions about when to take them, but the term "treatment" has very general application in experimental studies to represent any experimental manipulation. A key part of designing an experimental study is to decide on appropriate controls. For example, it's conceivable that the simple act of taking a pill daily might affect the occurrence or severity of colds, regardless of the contents of the pill. In other words, there might be psychological benefits of the treatment unrelated to the physiological effects of vitamin C itself. But the researchers were really interested in whether vitamin C *specifically* could protect against colds. Therefore, they established a control group that received placebo capsules instead of capsules containing vitamin C. More specifically, the vitamin C capsules contained 500 milligrams (mg) of vitamin C and 180 mg of lactose (milk sugar), whereas the placebo capsules contained 645 mg of lactose. The authors of the NIH study candidly state that the choice of lactose as the placebo was a hasty decision, necessitated by the desire to start the study within a few months of the time it was conceived. This turned out to be a fateful decision, as well as a hasty one.

Since Karlowski and his colleagues wanted to test both the prophylactic and therapeutic effects of vitamin C on the common cold, they divided their subjects into four groups. Members of each group were given two sets of capsules: maintenance capsules and supplemental capsules. They were instructed to take six of the former daily and six of the latter when they caught colds. For the first group, both types of capsules were placebos. For the second group, the maintenance capsules were placebos but the supplemental capsules contained vitamin C. The third group was the opposite of the second: vitamin C in maintenance capsules but placebos as supplemental capsules. The fourth group had vitamin C in both maintenance and supplemental capsules. For subjects receiving vitamin C in either maintenance or supplemental capsules, the dose was 3 grams (g) per day (6 capsules × 500 mg). For subjects who received vitamin C in both maintenance and supplemental capsules, the dose was 3 g/day when they did not have colds and 6 g/day when they did.

The most important feature of an experiment such as this is that subjects were *randomly* assigned to each of the four groups. This design is called a *randomized trial*, in contrast to the prospective and case-control designs described earlier. Randomization helps overcome the fundamental dilemma of nonexperimental designs—that other variables besides the one of interest may account for the results of a study. For example, suppose the NIH researchers had asked subjects to volunteer specifically to be in a group of their own choosing: the double-placebo group if they were skeptical of the hypothesis, the double-vitamin C group if they believed the hypothesis, or one of the intermediate groups if they weren't sure. The results of such an experiment would be impossible to interpret because any number of other factors might differ between these self-selected groups and be associated with different tendencies to choose the vitamin C treatment, as well as different susceptibilities to colds, thus compromising any interpretation of a beneficial effect of vitamin C. For instance, younger employees of NIH might be more skeptical of vitamin C than older ones, so more likely to select the placebo treatment, but younger employees might also be more susceptible to colds because frequency of colds typically decreases as people get older. So this "experiment" would show that those who took vitamin C got fewer, milder colds than those who took the placebo, a bogus conclusion.

The problem illustrated by this hypothetical example is essentially the same as that in the Swiss study of antioxidants and memory ability (Perrig et al. 1997). In that case, subjects differed in plasma levels of antioxidants because of individual differences in many potential factors: genetics, diet, age, sex, and lifestyle. We couldn't say with any assurance that differences in memory ability were due to differences in plasma levels of antioxidants or differences in one or more of these other confounding variables. In a randomized trial, the process of randomly assigning subjects to treatment and control groups helps alleviate this uncertainty because average values of potentially confounding variables are likely to be similar in treatment and control groups, especially with moderate to large sample sizes. This is a simple consequence of the randomization process. Imagine picking two softball teams from a pool of 100 men and 100 women. If we randomly select people for the two teams, they will probably have similar numbers of men and women. If, instead, we were picking a treatment and control group for an experiment, the same principle would apply.

I need to mention two other aspects of the design of this study that are common in medical and nutritional experiments before describing the results. First, this was ostensibly a double-blind study. This means that the subjects were not supposed to know whether they got vitamin C or placebo capsules for either their maintenance or supplemental supply, and the researchers were not supposed to know which treatment group subjects belonged to when they treated the subjects or recorded their symptoms. The purpose of the double-blind approach is to reduce the possibility that preconceptions of either subjects or researchers could bias results. If subjects knew they were taking vitamin C and believed in its efficacy, they might tend to downplay the symptoms of any colds they got. The same goes for researchers who are recording results. Second, the study relied partly on the subjects' own assessments of their health. They reported the number of colds they had and how long they were, whereas the researchers determined the severity of 20 different cold symptoms experienced by the subjects when they visited the clinic to get supplemental capsules containing either vitamin C or a placebo.

The average number of colds during the 9 months of the study was quite similar in the subjects taking a maintenance dose of 3 g of vitamin C and those taking a maintenance dose of placebo: 1.27 versus 1.36. This difference of 0.09 more colds per person in the placebo group was not statistically significant because the probability was greater than 50% that it could have been due to chance alone. One way to appreciate this is to notice that subjects with the same maintenance treatment but different supplemental treatments also differed by at least 0.09 in frequency of colds. This difference between supplemental treatments was 0.11 for those with the placebo maintenance treatment (the left pair of bars in Figure 2.2A) and 0.15 for those with the vitamin C maintenance treatment (the right pair of bars in Figure 2.2A). Yet supplemental capsules should have had no effect on the frequency of colds because they weren't given to the subjects until *after* they came down with a cold, and they were only given for 5 days so they should not have affected the likeli-

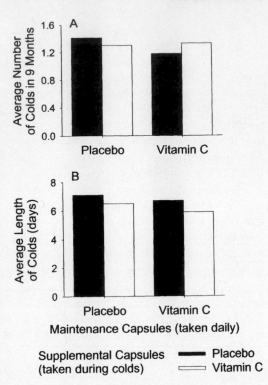

Figure 2.2. Frequency of colds (A) and average length of colds (B) for four groups of employees of the National Institutes of Health studied in an experiment by Karlowski and his colleagues (1975). Differences in the type of supplemental capsule taken (filled versus open bars) should not affect the frequency of colds but might affect their duration if vitamin C has a therapeutic effect. Differences in the type of maintenance capsule taken (the left and right pairs of bars) might affect both frequency and duration of colds.

hood of getting another cold weeks or months later. Therefore, these differences in cold frequencies associated with the supplemental treatment must have been due to chance, suggesting that the difference of similar magnitude between the maintenance treatments was probably also due to chance.

The story about potential therapeutic benefits of vitamin C is much more interesting. Karlowski's group (1975) found that colds lasted an average of 7.14 days for subjects in the double-placebo group (i.e., placebo as both their daily maintenance treatment and their supplemental treatment when they got colds), 6.59 days for subjects taking vitamin C either daily or as a supplement but not both, and 5.92 days for subjects taking vitamin C daily when they were well and as a supplement when they were ill (Figure 2.2B). These results suggest a small but significant benefit of vitamin C in reducing the length of colds, by about 0.5 days for an intake of 3 grams per day and by a little more than 1 day for an intake of 6 grams per day. In addition, several symptoms were less severe in subjects taking vitamin C than in those taking the placebo.

Perhaps you've already guessed the problem with these results. Because the NIH researchers organized the study hastily, they used a placebo that tastes sweet (lactose), whereas vitamin C tastes sour. The pills were provided in capsules, so this wouldn't be a problem if they were swallowed whole, but . . . over the nine months of the study, some participants evidently couldn't resist the temptation to bite into their capsules to try to determine which

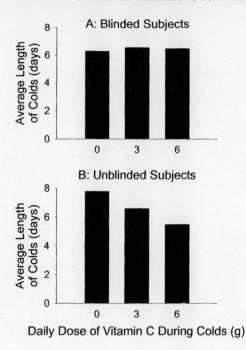

Figure 2.3. Average length of colds in subjects in the NIH study who did not guess their treatment group (A) and subjects who did guess their treatment group, whether successfully or not (B). The daily dose of vitamin C during colds was 0 for the double-placebo group, 3 grams for the groups that received placebo in maintenance capsules and vitamin C in supplemental capsules or vice versa, and 6 grams for the double–vitamin C group. Data from Karlowski et al. (1975).

group they belonged to. In other words, the double-blind was compromised. The researchers learned of this problem early in the study (some participants simply told them they had tasted the capsules and identified them). Therefore the participants were given a questionnaire at the end of the study that asked them to guess whether they had been taking vitamin C or the placebo. About 54% had guessed their daily maintenance treatment; 77% of these guesses were correct. About 40% had guessed their supplemental treatment; 60% of these guesses were correct.

When the NIH researchers examined the results separately for subjects who did not guess either their maintenance or supplemental treatment and subjects who guessed one or both of these treatments, a striking pattern emerged (Figure 2.3). For the nonguessers, there was no difference in the duration of colds between those who got a double-placebo treatment, a single dose of 3 g of vitamin C during their colds from either maintenance or supplemental capsules, or a double dose of 6 g of vitamin C during their colds (Figure 2.3A). For the guessers, there was a clear reduction in the length of colds for those receiving 3 g of vitamin C and a further reduction for those receiving 6 g (Figure 2.3B). For this analysis, the guessers, called "unblinded subjects" in Figure 2.3B, include those who guessed one or both of their treatments correctly and those who were wrong. This illustrates an apparent placebo effect: if a patient thinks he or she is getting a treatment, there is a psychological benefit comparable to any direct physiological benefit that the treatment may have. In this case, the results suggest that individuals who thought they were receiving a placebo had longer colds than those who thought they were receiving vitamin C, regardless of which treatment they were actually getting.

This study illustrates one of the major pitfalls of experimental research: the difficulty of setting up and maintaining a suitable control group. In this case, comparison of the vitamin C treatments with the placebo controls was complicated because some subjects guessed which group they were in. Karlowski and his colleagues concluded that a large daily dose of vitamin C did not prevent colds (Figure 2.2A) and that taking vitamin C during a cold did not shorten the cold (Figure 2.3A) or reduce the severity of cold symptoms (data not shown). These conclusions were based on analyzing a subset of their data—those for subjects who were truly in the dark about which of the treatment or control groups they belonged to. The researchers also found intriguing evidence for a placebo effect (Figure 2.3B), a problem that still bedevils medical research (Hróbjartsson and Gøtzsche 2001).

Despite its flaws, this 1975 study was well received by the medical establishment, probably because the authors were affiliated with the National Institutes of Health and the study was published in one of the premier medical journals in the United States. Nevertheless, the conclusions have been criticized by advocates of the beneficial effects of vitamin C. The most recent and most detailed critique was written by Harri Hemilä (1996), a scientist with the Department of Public Health at the University of Helsinki in Finland. Hemilä has been an avid booster of the vitamin C hypothesis in a series of articles published in the 1990s. All of these articles involve reviews of previous studies, in some cases with reanalyses of the original data. To illustrate Hemilä's approach, the title of his 1996 article in the *Journal of Clinical Epidemiology* in which he criticized the NIH research is "Vitamin C, the Placebo Effect, and the Common Cold: A Case Study of How Preconceptions Influence the Analysis of Results." In this article, Hemilä misrepresents some of the results of Karlowski's group and misinterprets other results, but he does raise one interesting issue. He suggests that the shorter and less severe colds of subjects who guessed correctly that they were getting vitamin C might be due to the fact that vitamin C really did reduce the duration and severity of colds, which in turn enabled the subjects to guess their treatment correctly. In other words, the placebo effect isn't the *cause* of an *artifactual* relationship between vitamin C intake and milder colds but rather a *consequence* of a *real* relationship between vitamin C and milder colds. This idea illustrates the complexities of disentangling cause-effect relationships in medicine, which will be explored further in Chapter 6. In theory, Hemilä's hypothesis could be tested by comparing characteristics of colds in subjects who were getting vitamin C and guessed correctly that they were getting vitamin C and in subjects who were not getting vitamin C but guessed that they were. The NIH researchers considered making this comparison in their original article, but didn't believe the sample sizes in these subgroups were sufficient.

The second author of the NIH study was Thomas Chalmers, who was director of the Clinical Center of NIH when the study was conducted. In introducing this study, I quoted Chalmers's expression of pride in the work. This quotation came from his one-page rebuttal to Hemilä's six-page critique. Chalmers concludes his rebuttal as follows: "In summary, I resent the time that I have had to devote to this author's biased defense of his late mentor's [Linus Pauling] infatuation with ascorbic acid. It may be that a properly done, unbiased, and updated meta-analysis[7] of the RCTs [randomized controlled trials] should be carried out, but I think it would be a waste of time" (1996:1085). It's rare to see such candid expression of emotion in the technical scientific literature, although strong feelings held by proponents of different hypotheses can sometimes be glimpsed at scientific meetings.

There have been numerous experimental tests of the effects of vitamin C on the common cold since the early 1970s. One consistent result is that taking large daily doses of vitamin C in an effort to prevent colds is futile. Even Hemilä agrees with this conclusion, although he thinks it may be possible that vitamin C has a prophylactic benefit for people who are physically stressed or suffer from borderline malnutrition. Experimental studies have provided more support for the hypothesis that taking high doses of vitamin C at the beginning of a cold reduces its duration and severity, although even in this case there is a lot of variation in results of various studies. In particular, Robert Douglas and his colleagues (2001) recently reported a suspicious pattern of greater apparent beneficial effects of therapeutic doses of vitamin C in more poorly designed studies. However, all of the experimental studies of vitamin C and the common cold have not yet been thoroughly and systematically reviewed, so we can't reach a definitive conclusion. Nevertheless, I'd like to briefly describe one of the most recent experimental studies to contrast some of its methods with those of the NIH study in the early 1970s. Then I'll make some concluding general points about the role of experiments on human volunteers in medical research.

Carmen Audera and three colleagues (2001) at the Australian National University (ANU) in Canberra studied the therapeutic effects of vitamin C on the length and severity of colds in the staff and students of ANU in 1998–1999. They solicited volunteers much as the NIH researchers did and used similar criteria for selecting subjects. Since Audera's group was interested specifically in testing the therapeutic effects of vitamin C, subjects were instructed to take medication only at the onset of a cold. Specifically, when they experienced two of several typical cold symptoms for at least 4 hours, or "four hours of certainty that a cold is coming on" (2001:360), they started taking the pills they had been given and did so for the first 3 days of the cold.

In this study, the placebo tablets contained 0.01 grams of vitamin C, and the dose was three tablets, or 0.03 g/day for 3 days. There were three additional treatment groups. One received 1 g of vitamin C per day for 3 days; one received 3 g of vitamin C per day; and one received BioC, containing 3 g of vitamin C per day plus four other substances thought to alleviate cold symptoms (e.g., rose-hip extract). The advantage of using a small dose of vitamin C in the placebo was that the taste of these tablets was apparently indistinguishable from that of the tablets for the three treatment groups that received much higher doses, yet the small dose was far below a level that ad-

vocates of vitamin C believe would be necessary to treat colds. Instead, this
small placebo dose was comparable to the minimum daily requirement of vitamin C to prevent scurvy. In fact, only 17% of the subjects guessed their
treatment group, and the majority of these guesses were incorrect. Contrast
this figure with the much higher percentage of subjects who guessed their
treatment group in the NIH study, in which lactose was used as the placebo.

As in the NIH study and in any true experiment, the subjects were randomly assigned to the four treatment groups. One problem of the study was
that a fairly large number of participants dropped out before reporting any
colds. The authors don't say whether dropouts were more likely to be students or staff at ANU. However, the numbers of colds for which data were
collected were comparable in the four treatment groups, suggesting that subjects in each of the four groups were equally likely to drop out of the study. A
second and more serious problem was that the subjects were responsible for
initiating their own treatment and for recording the severity of their cold
symptoms. How might this bias the results? If subjects in different treatment
groups were similarly accurate or inaccurate in recording their symptoms, reliance on the subjects themselves to record the data shouldn't produce systematic differences between the treatment groups. For example, if there was
a tendency to exaggerate symptoms, this should increase the average severity
score to the same extent in all four groups, so the differences among groups
should be the same as if the symptoms were not exaggerated. One of the purposes of random assignment of subjects to treatment groups in this study was
to minimize the likelihood that some groups had more hypochondriacs than
others, that is, to ensure that the average tendency to exaggerate symptoms
was similar in all four groups. However, the study might have been compromised if subjects didn't initiate self-treatment soon enough, even if there were
no differences in this lag time among groups. In fact, the average time between the beginning of cold symptoms and taking the first dose of medication was 13.4 hours, in contrast to the 4 hours specified in the instructions.
This average time didn't differ among groups, but a vitamin C advocate
could argue that a high dose needs to be taken at the very beginning of a cold
to be effective, so to have a fair test of the vitamin C hypothesis, the first dose
should be taken much sooner than 13 hours after the beginning of cold
symptoms.

Based on data for 184 colds in 149 subjects, Audera's group found no
therapeutic benefits of vitamin C. The average length of colds was actually
shortest for subjects in the placebo group, and the cumulative index of severity at 28 days after the cold started was second lowest in the placebo group
(Figure 2.4). However, there were no statistically significant differences
among groups in these summary measures or in any more specific measures
such as duration of nasal symptoms, throat symptoms, or systemic symptoms
(e.g., fever, headache, and achiness). As suggested in the previous paragraph,
this study was not foolproof, but it seems to provide fairly persuasive evidence against the hypothesis that large doses of vitamin C can be used successfully to treat the common cold. More important, comparing this study

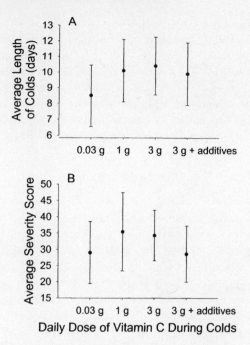

Daily Dose of Vitamin C During Colds

Figure 2.4. Average duration of cold symptoms (A) and average severity score (B) for subjects in four treatment groups in a study at Australian National University (Audera et al. 2001). The treatments were given during the first 3 days of colds. The dose of 0.03 grams per day represents the placebo; the dose of 3 grams plus additives represents "Bio C," which contained bioflavenoids, rutin, hisperidin, rose hip extract, and acerola in addition to 3 grams of vitamin C. Subjects rated the severity of cough, nasal, throat, and systemic symptoms on a scale of 1 to 3 each day; B shows the cumulative total of these scores after 28 days. This time period incorporates the full lengths of the longest colds. The vertical bars show 95% confidence intervals, a standard index of variability among individuals (see Chapter 8). For each group, the probability is 95% that the true average lies within the 95% confidence interval. The large overlaps in these confidence intervals implies that the differences among treatments are not significant.

with the NIH test of the same hypothesis illustrates many of the subtle problems that can arise in conducting nutritional or medical experiments with human volunteers.

While I was completing the first draft of this chapter, two articles about other treatments for the common cold appeared in my local newspaper on successive days. The first was a front-page story that announced "New Drug Could Be Common Cold Cure." The drug is called pleconaril and was tested in a randomized, controlled trial much like the experimental tests of vitamin C. The pharmaceutical company that developed the drug applied for approval from the U.S. Food and Drug Administration (FDA) to sell it, but in March 2002 an Advisory Committee to the FDA recommended that further studies be done before approval was granted.[8] Enough pleconaril to treat one cold would sell for about \$40, much more than the cost of vitamin C—if vitamin C were effective. The news report made no mention of all of the previous research on vitamin C but simply stated, "Scientists have developed the *first* [emphasis mine] medicine proven to reduce the length and severity of the common cold" (*Reno Gazette-Journal*, 18 December 2001). Pleconaril apparently works by directly attacking rhinovirus, the most common cause of the common cold. The second article touted the benefits of unfiltered beer for treating cold symptoms. This sounds to me like the best approach of all.

ADVANTAGES AND DISADVANTAGES OF OBSERVATIONAL
AND EXPERIMENTAL STUDIES

In this chapter, we've examined both nonexperimental and experimental studies to test various purported health benefits of vitamin C and other antioxidants. Although we've seen that experimental studies have pitfalls, I've emphasized their advantages, which make them the gold standard of medical research for good reason. However, we shouldn't simply dismiss the role of prospective studies, case-control comparisons, and other kinds of nonexperimental designs in medical research. These kinds of studies can provide essential information when experimental studies are impractical or unethical. For example, no one seriously doubts that smoking has multiple bad consequences for human health, yet much of the evidence for this belief comes from large-scale, long-term comparisons of smokers with matched groups of nonsmokers. In this case, matching subjects for as many potential confounding variables as possible has been a reasonable substitute for randomly assigning "volunteers" to smoke or refrain from smoking for several years. Indictment of smoking as the most significant controllable health risk has been reinforced by the fact that many different studies of diverse groups of people have all pointed to the same conclusion. Furthermore, experimental and nonexperimental studies often provide complementary kinds of evidence to answer questions not only in medicine but in other biological sciences as well. I'll illustrate the process of marshaling multiple types of evidence to answer biological questions in Chapter 4, using two recent ecological examples.

Unfortunately, the culture of science sometimes leads to excessive valuation of experimental methods. There is a fascinating example of this in the case of a disease of newborns called persistent pulmonary hypertension. With conventional medical treatment (CT), only 20% of patients survived this disease. In the late 1970s, a group at the University of Michigan (Bartlett et al. 2000) began testing a new treatment called extracorporeal membraneous oxygenation (ECMO) in which the patient's blood is passed through a heart-lung machine outside the body for several days. They were able to increase survival rate to 80%, but weren't completely confident of this seemingly dramatic success. Perhaps their patients differed in some unknown way from the earlier group of patients, and this difference rather than their new treatment accounted for the difference in survival. Therefore, they designed a small randomized trial to compare ECMO to CT. Because the new treatment seemed so successful and they wanted to make it available to as many patients as possible, they used a randomization method called "randomized play the winner." The treatment for the first baby suffering from persistent pulmonary hypertension was to be determined completely randomly; that is, there would be 50% probability that it would receive CT and 50% probability that it would receive the new therapy. If the treatment was successful, its probability would be increased for the next baby. As it happened, the first baby was assigned the new ECMO treatment and survived whereas the second baby was assigned CT and died. By this time, according to the prespecified protocol for randomization, the probability of assigning a baby to the new treatment was 75% and to the old treatment was 25%. The next 10 babies received the new treatment and all survived; the results of the study were reported in 1985 as 11 survivors of the 11 babies treated by ECMO and 1 death of the one baby treated by CT.

This study was severely criticized by other researchers because only one patient received CT, so a follow-up study with a standard randomization scheme similar to that used in the vitamin C studies was designed and carried out (Ware 1989; Royall 1991). There were nine patients in the ECMO treatment group, all of whom survived, and 10 patients in the control group that received CT, six of whom survived. This example raises a wrenching ethical dilemma. Do you think the deaths of four infants who received CT in this final randomized trial were "necessary" in some sense to demonstrate conclusively the value of the new treatment? Or do you think the medical community should have been satisfied with the initial comparative data in which the control group was not randomly selected but was simply made up of patients born before the new treatment was developed? How would your answers differ if the benefits of a new treatment weren't so dramatic? Seemingly dry and technical aspects of scientific methodology sometimes have profound practical and ethical implications.

RECOMMENDED RESOURCES

Centre for Evidence-Based Medicine. 2003. Evidence-based medicine. http://www.cebm.net (accessed October 1, 2003). This is a web site based in Great Britain which promotes and supports evidence-based medicine, that is, the use of rigorous standards of evidence in evaluating medical procedures.

Giere, R. N. 1997. *Understanding scientific reasoning*, 4th ed. Harcourt Brace College Publishers, Fort Worth, Tex. A basic textbook on how scientists test hypotheses and evaluate evidence with lots of good examples.

Ware, J. H. 1989. Investigating therapies of potentially great benefit: ECMO (with discussion). *Statistical Science* 4:298–340. Advocates of many different positions about ethical aspects of experimentation in medicine exchange views.

Chapter 3

Can Police Dogs Identify Criminal Suspects by Smell?

Using Experiments to Test Hypotheses about Animal Behavior

We know from our everyday experience with pets that animals have different sensory abilities than humans. In many cases, it's obvious that these sensory abilities far exceed our own. Even the most casual observers, for example, would soon realize that their dogs live in a world dominated by odors and that they can detect and distinguish odors unknown to the owners. Indeed, some of the most fascinating stories in biology involve the discovery of specialized sensory abilities in particular types of animals that are totally lacking in humans. A classic example is Donald Griffin's (1986) discovery of echolocation by bats. Many bats emit streams of high-frequency sounds, well above the highest frequency that we can hear. These sounds bounce off objects in the environment, and the bats use the resulting pattern of echoes to navigate under pitch-black conditions at night. They also use echolocation to locate their prey, such as moths and other flying insects. This is an exquisite adaptation for success for a highly mobile, nocturnal, aerial predator of small prey that are also highly mobile.

The excitement of Griffin's discovery arises from the fact that, as humans, our primary tools for learning about the world are our own sensory abilities of taste, touch, smell, and especially hearing and vision. This has always been the case and is still largely true, despite the complex machinery for collecting and analyzing data that we associate with modern science (after all, the output of the machines generally has to be looked at or listened to by people in order to be interpreted). Because we depend on our limited sensory abilities for doing science, learning about sensory abilities of other animals that differ qualitatively or quantitatively from our own is particularly challenging. In the case of echolocation, European scientists discovered in the late 1700s that blinded bats could navigate in a room but bats whose ears were plugged with wax could not. Griffin repeated these experiments in the 1940s but was able to use new acoustic equipment to record the high-frequency sounds of the bats, finally solving the mystery of how bats navigate that had intrigued the European researchers 150 years earlier.

Donald Griffin (1986) describes his discovery of echolocation in a wonderful book for a general audience called *Listening in the Dark*. It's a tale of imagination, invention, and the design and execution of critical experiments. I'll use an example that is somewhat less exotic but especially well suited to illustrating the experimental method. It is the opposite of the echolocation story because it is about the *limits* of the olfactory abilities of dogs, which are often assumed to be virtually unlimited . Rather than inspiring our awe that animals can do something that we never imagined would be possible, such as navigate by echolocation, this story about the sense of smell in dogs shows how common knowledge can sometimes get ahead of scientific evidence, with significant practical consequences. In debunking the health benefits of vitamin C in Chapter 2 and the olfactory abilities of dogs in this chapter, I don't want to leave you with the impression that experiments are always used to discredit popular hypotheses. The next chapter will illustrate the positive role of experiments combined with other kinds of evidence to evaluate hypotheses.

THE MYTHIC INFALLIBILITY OF THE DOG

Humans have appreciated the olfactory abilities of dogs since antiquity and have developed various breeds to capitalize on these abilities. The use of dogs to assist in hunting is the oldest and probably most familiar example, but trained dogs are also used in various law enforcement tasks such as tracking fugitives and sniffing out narcotics. For example, James Earl Ray, who was convicted of killing Martin Luther King, Jr., escaped from the Brushy Mountain State Penitentiary in Tennessee but was tracked by two bloodhounds and quickly returned to prison. Dogs are also used by the police to identify suspects in lineups. Just as human witnesses to a crime may be asked to select a matching suspect from a lineup, a trained dog may be given a tool or piece of clothing from a crime scene to smell and then be presented with a lineup that includes a suspect and several other people. If the dog shows by its behavior that the odor of the suspect matches that of the object from the crime scene, testimony to this effect by the dog's handler is typically admissible in a trial. In fact, Andrew Taslitz reported in 1990 that thousands of these lineups have been done in the United States since 1920, contributing to convictions for robbery, rape, and murder and sentences up to life imprisonment or death.

What is the legal basis for admitting evidence from scent lineups in trials? According to Taslitz, this basis is remarkably weak, at least in the United States. Most courts use four criteria, which were developed for deciding the validity of tracking and simply transplanted to scent lineups: that the dogs should belong to a breed "characterized by acuteness of scent and the power to discriminate among individual human beings" (1990:120), that they be trained for tracking, that there be evidence of their reliability in tracking, and that there be other evidence independent of the dogs that is consistent with the scent identification. In many cases, convictions have been based primarily on scent identification by trained dogs, with additional evidence being only circumstantial. In practice, courts have relied almost exclusively on the claims of handlers about the abilities of their dogs. For example, in one robbery case, a handler testified that a tracking dog named Bobby was 100% accurate in training, as well as in four previous criminal cases. Both the original court and an appeals court accepted this testimony as "ample proof of reliability to justify admitting the results of Bobby's tracking" (1990:55).

Taslitz suggests that the faith of judges and juries in scent identification by dogs is rooted in a kind of mystical belief in man's best friend. Our culture has numerous legends, as well as true stories, attesting to the loyalty of dogs. In Homer's *Odyssey*, for example, Odysseus returns home after 20 years and is recognized by his aged dog Argus, who dies in the process of greeting his master. Argus probably recognized Odysseus after this long absence at least partly by smell. Because dogs are known for their loyalty, honesty, and integrity, we tend to accept evidence based on scent identification by dogs relatively uncritically. By contrast, imagine using *cats*, with their reputation for deviousness, in this way.

Because of "the mythic infallibility of the dog," in Taslitz's (1990:20) words, defense lawyers have so far been unsuccessful in persuading courts to apply the Frye Rule to evidence from tracking or scent lineups. The Frye Rule states that a scientific principle should be well established and generally accepted by scientists in the appropriate field for expert testimony based on that principle to be used as evidence in court. The purpose of the Frye Rule is to prevent juries from being swayed by the testimony of experts when the scientific foundation for that testimony is weak or nonexistent. In several cases, defense attorneys have argued that convictions should be overturned because the Frye Rule was not applied to the testimony of dog handlers. Judges have given various reasons for denying this argument, including the claim that such identification is not based on science, so the Frye Rule doesn't apply. This is clearly faulty logic. The olfactory abilities of dogs are subject to experimental testing, and a few such experiments have in fact been done (Brisbin et al. 2000). What is the evidence that trained dogs can recognize unique odors of individual people and use this ability to accurately identify subjects in lineups?[1]

In an interesting study with twins, Peter Hepper (1988) tested the hypothesis that humans have individual odors that can be recognized by dogs. Hepper was particularly interested in the role of genetic and environmental factors in causing people to have different odors. Therefore, he used three sets of twins in his experiments. The first set consisted of male fraternal twins that were 2 to 3 months old. These twins were genetically different but had a common environment because they were being raised in the same home. In particular, they ate the same foods, so any effect of diet on body odor should have been the same for both members of a pair. The second set consisted of male identical twins[2] that were 2 to 3 months old. These twins were not only genetically identical but also shared a common environment. The third set consisted of male identical twins that were between 34 and 50 years old, lived separately, and ate different foods. These twins were genetically identical but had different environments.

Four dogs were thoroughly trained in scent discrimination before being used in the experiments. This doesn't mean that the dogs learned *how* to distinguish between two similar scents; if they had an ability to discriminate, it was probably innate. Instead it means that the dogs were trained to show by their behavior that they were making a choice. The basic protocol was to wash each of the subjects with the same soap, then carefully rinse off the soap (presumably the adult subjects washed themselves). Four T-shirts for each pair of twins were also washed with the same detergent. Each twin then wore one T-shirt for 24 hours and a second T-shirt for the next 24 hours. Finally, a dog was presented with a T-shirt scented by one person, then given a choice between the second T-shirt worn by that person and one of the T-shirts worn by his twin. The dog's handler did not know which was the correct match.

The results of this experiment suggest that dogs can use *either* genetic or environmental factors to discriminate between people. The average percentage of correct choices was 89% for the infant fraternal twins sharing a common environment, 49% for the infant identical twins sharing a common environment, and 84% for the adult identical twins living in different environments. Since the dogs were choosing between two T-shirts in each trial, they should have been able to get 50% of their choices correct just by chance. The 49% rate of success at distinguishing twins with the same genes and same environment is not different from chance performance, suggesting that the dogs got no useful cues for discrimination in this situation. This provides a baseline for comparison with the other two types of trials, where performance of the dogs was significantly better than chance. For the trials with adult identical twins, environmental differences such as diet must have been the basis for discrimination by the dogs. For the trials with infant fraternal twins, genetic differences must have been the basis for discrimination because infant identical twins could not be distinguished.[3]

Are these results sufficient to validate the use of dogs to identify suspects in scent lineups? Not by a long shot, for at least four reasons (not including the fact that 2-month-olds rarely commit crimes). First and least important, Hepper tested only four dogs, so we can't say if their abilities are common or unusual among dogs in general (this is least important because Hepper did show that *some* dogs have remarkable abilities to discriminate human odors). Second, the success rates were impressive, but the performance of the dogs was not perfect even after intensive training. Dogs made mistakes in identification 11% of the time in trials with infant fraternal twins and 16% of the time in trials with adult identical twins. Error rates of this magnitude that resulted in convictions of innocent people would be unacceptable. Third, dogs used in forensics often have to compare an odor from one body part of a perpetrator (e.g., head odor on a hat left at the crime scene) with an odor from another body part of a suspect in a lineup (the suspect and several control individuals use their hands to apply scent to a test object such as a metal cylinder). In Hepper's experiments, odors all came from the torsos of the twins. Just as individuals may differ in smell, various body parts of the same individual may also differ. Can dogs be trained to ignore this anatomical variation and detect a component of the olfactory signal that's common to all body parts of an individual, if such a common signal exists? Finally, Hepper's experiment differs from real-world forensic practice in that a suspect in a lineup may not be the same as the perpetrator of a crime. If this is the case, a dog given an object from a crime scene to sniff should pick no one from the lineup because no odors of these people would match the smell of the object from the crime scene. In Hepper's experiment, by contrast, one of the T-shirts presented to the dog was always a correct choice.

I. Lehr Brisbin and Steven Austad (1991) did a small experiment to see how the ability of dogs to distinguish between the scents of two different individuals was affected by the body parts that supplied the scents. They used three dogs and modeled their procedure after that used in competitions authorized by the American Kennel Club. In these competitions, dogs are required to select metal and leather dumbbells scented by their handlers' hands when given a choice between these and dumbbells scented by the judge's hand. This differs from forensic practice in an important way because we might expect that it would be easier for dogs to distinguish between the familiar odors of their handlers and the unfamiliar odors of judges than between the odors of various people in lineups, which would all be equally unfamiliar. This means that the experiment of Brisbin and Austad gives the benefit of the doubt to the hypothesis that dogs can generalize across body parts to identify individuals. If their dogs were able to do this, it might be because the handlers' odors were so familiar, which would provide only weak evidence that dogs could generalize in scent lineups. However, if the dogs used by Brisbin and Austad were not able to generalize, it seems unlikely that dogs could do so in the more challenging circumstances of the lineups.

The dogs were trained by using standard guidelines of the American Kennel Club for obedience training of "utility dogs" for at least 6 months, then further trained in the specific procedures used by Brisbin and Austad in their experiments. In each trial, two scented dumbbells were placed on a board about 10 feet from the dog and handler while the dog and handler were facing in the opposite direction. When the dumbbells were in place, the dog and handler were instructed to turn around to face the board. The handler gave the dog a command to go to the board. Sniffs of each of the dumbbells were recorded, as well as which dumbbell was retrieved. The experimenters positioned the dumbbells in such a way that the handler didn't know which was the correct choice.

The dumbbells were scented by having a person hold a dumbbell in his or her hand for 30 seconds or using tongs to position the dumbbell in the interior crook of the elbow and having the person hold it tightly there for 30 seconds. Several types of trials were conducted. First, the dogs had to choose between (1) a dumbbell scented by the handler's hand and one with no human scent or (2) a dumbbell scented by the handler's elbow and one with no human scent. The dogs had no trouble with these discriminations, averaging 93% correct in the first case and 86% correct in the second case. The dogs also had little trouble in the second type of trial, distinguishing a dumbbell scented by the handler's hand from one scented by a stranger's hand, which is comparable to the task used in obedience competitions. Success rates of the three dogs were 69%, 70%, and 90%, for an average of 76%. However, when given a choice between a dumbbell scented by their handler's elbow versus one scented by a stranger's hand, the dogs were less successful. Success rates were 70%, 57%, and 46%, and the average of 58% was not significantly different from the chance performance of 50%. Since the dogs were trained to identify the hand odors of their handlers, they may have been confused by differences between elbow odors of their handlers and hand odors of strangers. It is interesting that they could distinguish elbow and hand odors of their handlers, with an average success rate of 78%.

In a nutshell, these results suggest that different body parts of the same person have different odors. This shouldn't be surprising, although the fact that the elbow and hand of the same arm smell differently may be a bit curious. More important, the results suggest that dogs do not automatically generalize from one body part of an individual to another to discriminate between two people with individually distinctive odors, *if the dogs are trained by using standard methods*. It's quite conceivable that a training regime could be devised to improve the performance of dogs in this task, but present methods that are used not only to train dogs for competitions sponsored by the American Kennel Club but also for scent identification in police work are not adequate. This undermines one of the fundamental assumptions of the use of evidence from scent lineups in court.

A group of researchers led by Ray Settle of the Police Dog Training School in Preston, England, developed and tested a training routine that they thought might be more effective than standard methods of training dogs to generalize across body parts in identifying individuals (Settle et al. 1994). Settle's group used seven dogs of various breeds. They collected body scents from more than 700 volunteers from various schools, a local business, and a nursing home. Each volunteer was given four pieces of cotton cloth that had been washed and placed in a glass jar. The volunteer was asked to place each piece of cloth on a different part of his or her body for 30 minutes, then replace the scented cloths in the jar. The choice of body parts to be scented was up to the volunteers.

The dogs were tested by giving them one cloth from one of the volunteers to sniff and then either presenting them with a group containing another cloth scented by the same volunteer plus five cloths scented by five other volunteers or presenting them with six steel tubes that had been held by volunteers for 5 minutes. These procedures are similar to those used in actual scent lineups, at least in Europe. Handlers trained their dogs in a series of progressively more difficult discriminations. For example, in an early step in training, the dogs had to identify a scented cloth when it was placed in the training room with five others that had been washed and handled with tongs, so they had no human scent. In successive steps, the dogs had to discriminate a target scent from one other human scent, then two others, and so on. Training lasted for 9 months.

After this regimen, the dogs were correct 80% of the time on average in the first type of test, in which they had to discriminate among six cloths scented by different people, and 85% of the time in the second type of test, in which they had to discriminate among six steel tubes handled for 5 minutes by different people. In each case, since one of the six choices matched the target scent, we would expect the dogs to be correct one-sixth of the time (17%) purely by chance. Accuracy of 80 to 85% is much better than chance performance, implying that the dogs really had learned to distinguish individual human odors in a situation similar to that used in actual police lineups. Furthermore, Settle and his colleagues suggested that the dogs were identifying individuals regardless of whether or not the body part used as a source of the target scent was the same as that used in the lineups.

These results of Settle's group are inconclusive, however, because the volunteers who provided cloths scented by four different parts of their bodies handled all of the cloths and placed them together in a closed glass container that was returned to the experimenters. In handling the cloths, hand odors may have been transferred to them; in keeping them in closed containers for up to 4 days, odors may have been transferred among them. This means that the odors on four cloths prepared by the same volunteer were probably more similar to each other than, for example, the odors on the dumbbells held in the hand and the elbow by a subject in the experiment of Brisbin and Austad (1991). Because of this likely odor contamination in the study by Settle and his colleagues, it may have been easier for the dogs to match a target scent to one of the scents in the lineup than if it had come from a distinctly different part of the body than the part used to create the lineup. Therefore, this study doesn't restore much faith in the validity of evidence from scent lineups, because it doesn't convincingly show that dogs can generalize from one body part to another to distinguish odors from different people, even though the training used by Settle's group was more extensive than that normally given to police dogs.

The most extensive and most recent set of experiments dealing with scent identification by dogs in police lineups was done by Gertrud Schoon for her dissertation at the University of Leiden in the Netherlands.[4] She is affiliated with the Department of Criminalistics and Forensic Science and the Ethology Group of the Institute of Evolutionary and Ecological Sciences at the university, indicating that she has broad training and diverse interests. Real progress in science often comes from people who bring a new perspective to a long-standing problem. In this case, Schoon's background in ethology (animal behavior) may have helped her to find creative ways of dealing with some of the practical problems of forensic science that we've been discussing.

Schoon (1996, 1997, 1998) examined several aspects of scent identification to find ways to improve the training of dogs and the operation of scent lineups, but I will discuss just one part of her work that addresses one of the most important pitfalls of these lineups. This is the possibility that a dog will falsely accuse a suspect by selecting the suspect's scent from a lineup when this scent does not match the target scent from an object left at a crime scene. Unlike all the experiments described so far that used a *match-to-sample* design, in which an odor matching the sample was always present in the array with which the dogs were tested, in the complicated real world of police investigations innocent suspects are sometimes arrested. In this case, the correct choice of a dog would be to select none of the odors in a lineup because none would match that from the crime scene. But there are various factors that might work against dogs making this response. They might select the odor that was *closest* to the odor on the target object, even if it wasn't identical. In fact, when human witnesses to crimes are shown pictures of several potential subjects, they tend to pick one that looks most similar to the person they saw at a crime scene, even if that person was not actually there (Wells et al. 2000). In scent lineups, the control scents often come from police officers who may be familiar to the dog, so the dog may pick the odor from the lineup that is least familiar, regardless of whether it matches the scent of the target object. The handler may believe the suspect is guilty and therefore reward the dog for making any selection at all. If the lineup consists of a suspect and several control individuals, none of whom are known to the handler, the handler may pick out the likely suspect by using visual cues and communicate this identification to the dog unconsciously.[5] Even if the lineup consists of objects like metal cylinders that were scented by using standardized methods, the handler, who is likely to be a police officer familiar with the case, may believe the suspect is guilty and therefore unconsciously encourage the dog to choose one of the cylinders when the correct response if the suspect was really innocent would be no choice. To date, Schoon is the only researcher who has rigorously tested the possibility of false identification of suspects based on dogs and scent lineups.

Schoon did her experiments with six tracking dogs that were trained for police work in the Netherlands. Each dog was used in 10 trials. In each one, a target scent was prepared by having a volunteer police employee treat an object as if it had been found at a crime scene. These objects were screwdrivers, wrenches, pistol buttplates, sweatshirt cuffs, and scent samples from the seat of the volunteer's car. In preparing these target scents, the volunteers were acting as if they were the perpetrators of a crime.

To prepare odors for scent lineups, Schoon gave volunteers two glass jars with six stainless steel cylinders in each. The volunteers were instructed to handle the cylinders in each jar for 5 minutes. In addition, one of the volunteers handled a piece of polyvinyl chloride (PVC) tubing, providing a "check" scent that was used as described below to be sure the dogs didn't have an inherent preference for one of the odors in the lineup but were really making a choice based on matching an odor in the lineup to a target odor. For each trial with a dog, one volunteer was designated as a suspect. In half of the trials with each dog, this "suspect" was the same as the "perpetrator" who had scented the objects from a hypothetical crime scene; in the remaining trials, the suspect and perpetrator were different people.

Each trial consisted of four tests of discrimination in which seven scented cylinders were laid out for a dog. One cylinder had the scent of the suspect, one cylinder had the scent of the person who provided the check scent, and the other five cylinders had scents of five other volunteers that Schoon called decoys. In the first two trials, the dog was allowed to smell the check scent on the PVC tubing, then shown the lineup of seven cylinders. In this case, the correct choice was to select the cylinder scented by the person who made the check scent. If the dog made the wrong choice in either of these tests, it was disqualified for that trial. For example, the dog might be disinterested in working that day and make no choice, or it might select the suspect's scent, indicating a preference for that scent even though it didn't match the check scent. In these cases, Schoon believed it was invalid to test the dog's ability to compare a perpetrator's and suspect's scent. One of her pragmatic suggestions is that police departments begin using this kind of performance check in actual lineups.

After these two trials, the check scent was removed from the lineup and the dog was given the scented object made by the person pretending to be the perpetrator. Then it was given two tests with six scented cylinders—one scented by the simulated suspect and five scented by decoys. Recall that in half of the trials with each of the six dogs, the suspect was the same as the perpetrator, whereas in the other half the suspect and perpetrator were different, and the handler did not know which trial was which. Since there were five types of objects, there were 10 trials per dog, for a total of 60 trials.

The performance of the dogs in this experiment was not particularly impressive. In 30 of the 60 trials, the dogs were disqualified because they made errors in the check tests. In the remaining trials in which the suspect was the same as the perpetrator, the dogs correctly selected the scent of the suspect in four trials, they selected a decoy's scent in five trials, and they made no selection in two trials, for a success rate of 4/11, or 36%. In the trials in which the suspect was not the same as the perpetrator, the dogs correctly made no selection in nine cases, they selected the suspect's scent in one case, and they selected a decoy's scent in nine cases, for a success rate of 9/19, or 47%. It's interesting that the dogs made some choice in a majority of these trials, suggesting that it may have been difficult for them to resist picking one item from the lineup even when there was no match to the scent on the object from the hypothetical crime scene. However, the effective error rate was only 1/19, or 5%, because in nine cases the dogs selected a decoy, who would presumably be known to be innocent in a real lineup. These estimates assume that the conditions of this experiment with simulated suspects and perpetrators are representative of actual forensic practice. Also, the estimates are not very precise because they are based on a small number of successful trials with only six dogs.

INTERPRETING EVIDENCE

Ignoring the limitations of its small sample size, what can we learn from this study? There are two categories of errors that can occur in methods used to match suspects to evidence from a crime scene, whether this evidence be fingerprints, tissue samples containing DNA, eyewitness accounts, or odors on objects identified by dogs. These errors are false positive identification, in which an innocent suspect is convicted because of an incorrect match between the evidence and the suspect, and false negative identification, in which a guilty suspect is acquitted because of failure to make a match between the evidence and the suspect. The general principle of jurisprudence, that a person is considered innocent until proven guilty beyond a reasonable doubt, implies that the first type of error is considered more serious than the second, at least in contemporary Western society. McCauliff (1982) surveyed 171 judges and found that the most common interpretation of "beyond a reasonable doubt" was that the chance of false positive identification was 10%. This means that innocent suspects might be convicted as frequently as 10% of the time, or, said another way, that conviction is reasonable if the likelihood of guilt is greater than 90%. Of course, in our legal system of trial by jury, the ultimate determinant of the meaning of "beyond a reasonable doubt" is the collective opinion of the jurors.

Before applying these ideas to the problem of scent identification, I'd like to discuss a seemingly different but in fact perfectly parallel problem, which is easier to analyze because fewer assumptions are necessary. Consider a diagnostic test for a disease, such as the occult-stool test for colorectal cancer in which blood in the stool is used as an indication that a person might have cancer in the lower portion of the digestive tract (Hoffrage et al. 2000). We can summarize the values needed for our calculations in a 2 × 2 table in which the two columns represent people with and without the disease and the two rows represent positive and negative results of the test (Table 3.1). The sensitivity of the test is 50%; that is, if a person has the disease, the chance of a positive test result is 50%. This means that the occult-stool test misses 50% of the cases of colorectal cancer. These are false negatives, represented by the value of 0.5 in the lower-left cell of the table.

The upper-right cell of the table shows the probability of a false positive test result: an occult-stool reading that indicates colorectal cancer in a person who does not actually have the disease. This probability is only 3%. In other words, for a person who is not afflicted with colorectal cancer, the chances are 97% that the occult-stool test would correctly be negative. But the most important issue is how to interpret a positive test result for someone whose

Table 3.1. The occult-stool test for colorectal cancer.

		Disease Status of Individual	
		Has Colorectal Cancer	Does Not Have Colorectal Cancer
Results of Occult-Stool Test	Positive	0.5	0.03
	Negative	0.5	0.97

Values in the first column are probabilities of positive and negative test results for people with colorectal cancer; values in the second column are probabilities of positive and negative test results for people who do not have colorectal cancer (Hoffrage et al. 2000).

disease status is not known. Stated more personally, if your doctor did an occult-stool test and reported a positive result, what are the chances that you have colorectal cancer?

Based on the top row of Table 3.1, you might assume that your chances of having colorectal cancer in this case are pretty high because the probability of a positive test result in a person with colorectal cancer is much larger than the probability of a false positive (0.5 versus 0.03). However, the significance of this depends on the frequency of colorectal cancer in the general population, which is only about 0.3% (Hoffrage et al. 2000). Why is this important? The easiest way to see how this information affects the analysis is to imagine applying the probabilities in Table 3.1 to a large, hypothetical group of people. Suppose we consider a group of 10,000. If this group is representative of the general population, 30 would have colorectal cancer. Of these 30, 15 would have a positive occult-stool test and 15 would have a negative test, based on the first column of Table 3.1. Of the remaining 9,970 people who do not have colorectal cancer, we expect there to be about 299 who would have false positive tests (0.03 × 9970). These numbers are shown in Table 3.2. Looking at the first row of Table 3.2, we see that the expected total number

Table 3.2. Test results and disease status of 10,000 hypothetical individuals, where the frequency of colorectal cancer is 0.3%.

		Disease Status of Individuals		
		Have Colorectal Cancer	Do Not Have Colorectal Cancer	Totals
Results of Occult-Stool Test	Positive	15	299	314
	Negative	15	9,671	9,686
Totals		30	9,970	10,000

I constructed this table by first subdividing the total of 10,000 individuals into 30 with colorectal cancer and 9,970 without colorectal cancer because 0.3% of 10,000 = 30 and 10,000 − 30 = 9,970. For those with colorectal cancer, half would be expected to have a positive occult-stool test and half would be expected to have a negative test based on Table 3.1, producing the values of 15 in the first numerical column here. For those without colorectal cancer, 3% would be expected to have a positive occult-stool test based on the upper-right cell in Table 3.1, and 3% of 9,970 = 299. Summing across the first row gives 314 positive tests among 10,000 people. Only 15 of these positive tests (5% of 314) would occur in people who actually have colorectal cancer.

of positive test results is 314, of which only 15 are actually associated with cancer. That is, only about 5% of people with a positive occult-stool test would actually have colorectal cancer, and you might be fortunate enough to be in the 95% without cancer, despite a positive test. Even though the likelihood of a false positive result for an individual without colorectal cancer is relatively small (3%) compared to the likelihood of a correct positive result for an individual with cancer (50%), the number of people without cancer is so much larger than the number with cancer that most positive test results occur in people without cancer.

Now let's return to scent identification by dogs. The results of Schoon's research are summarized in Table 3.3, which has the same format as Table 3.1. The critical question is similar to that for diagnostic testing for colorectal cancer: if a dog identifies a suspect as having an odor matching that from a crime scene, what is the probability that the suspect is guilty? Just as in the cancer example, we need more information than the values shown in Table 3.3 to answer this question. In the cancer example, the additional information was the frequency of colorectal cancer in the population. The parallel information for the scent-identification example would be the number of potential suspects for the crime. If we assume that only one person committed the crime and that there are 10 possible suspects, the proportion of suspects that is guilty is 1/10, or 10%, just as the proportion of people with colorectal cancer is 30/10,000, or 0.3%. Unfortunately, for most crimes it's not very clear how many potential suspects there are. However, imagine one of those classic murder mysteries on an estate in the English countryside. The owner, an eccentric bachelor, has 10 servants (the gardener, the butler, the cook, etc.). He invites 10 guests for a weekend of hunting. On Saturday evening, the owner is discovered murdered. A handkerchief with no identifying marks has been left on the floor of the library where the owner's body is found. The local constable brings his trained dog to match the scent of the handkerchief to that of one of the suspects. It's obvious that one of the 10 servants or 10 guests committed the dastardly deed, so the calculations can be made just as in the cancer example (to keep things reasonably simple, we'll assume that

Table 3.3. Scent identification by dogs in lineups based on Gertrud Schoon's (1998) research.

		Status of Suspect	
		Guilty (Suspect = Perpetrator)	Innocent (Suspect ≠ Perpetrator)
Identification by Dog	Positive (suspect selected)	0.36	0.05
	Negative (suspect not selected)	0.64	0.95

Values represent the proportion of trials in which the dogs made correct choices (the upper-left and lower-right cells), in which they failed to identify a guilty suspect (false negatives in the lower-left cell), and in which they incorrectly identified an innocent suspect (false positives in the upper-right cell).

there was only one murderer). These calculations are shown in Table 3.4. The bottom row shows the 20 total suspects divided into two groups, one guilty person and 19 innocent ones. The first column of Table 3.3 remains the same in Table 3.4 because we assume there is only one guilty person. But the values in the second column of Table 3.3 are increased in Table 3.4 because there are 19 innocent suspects who might be falsely identified by the dog. Specifically, the values in the second column of Table 3.3 are multiplied by 19 to get the second column of Table 3.4. This implies that if the dog selects a suspect (the top row of Table 3.4), the chance that the suspect is guilty is only 27% (0.36/1.21 = 0.27). Even though the probability of a false positive (0.05) is much lower than the probability of correctly picking the guilty person (0.36), there are so many more opportunities for the dog to select an innocent suspect than the guilty person (19 versus 1) that the likelihood that it will pick an innocent person is 73%.

Most criminal cases in which a scent lineup might be used aren't as straightforward as this because the total number of possible suspects isn't known. However, there is often other evidence that links a particular suspect to a crime. At least in theory, this other evidence can be used to estimate the likelihood that the suspect is guilty, independently of whether a dog selects the suspect's scent from a lineup. This is called the *prior probability* of guilt; that is, the probability of guilt before taking into account the results of the scent lineup. In the example of the English murder mystery, the prior probability of guilt is 1/20, or 5%, for each suspect because there are exactly 20 suspects and we have no other evidence pointing toward any one of the 20. This prior probability is represented in the last row of Table 3.4. In the scent lineup, the constable's dog matches the scent of one of the suspects to the odor on the handkerchief left by the victim's body. Because trained dogs really do have some ability to identify individual humans by smell, this increases the likelihood that the suspect identified by the dog is guilty. After the lineup, the so-

Table 3.4. An application of Schoon's (1998) results on accuracy of scent identification by dogs.

| | | Status of Suspect | | |
		Guilty (Suspect = Perpetrator)	Innocent (Suspect ≠ Perpetrator)	Totals
Identification by Dog	Positive (suspect selected)	$0.36 \times 1 = 0.36$	$0.05 \times 19 = 0.95$	1.31
	Negative (suspect not selected)	$0.64 \times 1 = 0.64$	$0.95 \times 19 = 18.05$	18.69
Totals		1	19	20

We assume that there are 20 possible suspects, 10 servants and 10 guests, only one of whom is guilty of murdering the owner of an English country estate. This provides the totals in the bottom row. The totals for each column are multiplied by the appropriate probabilities in each cell of Table 3.3 to get the values in this table, which can be used to compute the probability that an innocent suspect is mistakenly identified as guilty. This probability is 0.95/1.31 = 0.73.

called *posterior probability* of guilt is 27%. This is a substantial increase over the 10% prior probability of guilt, but it is far short of the standard expressed by the phrase "beyond a reasonable doubt."[6]

Suppose we wanted to be 90% sure of blaming the correct person for the crime. How many suspects would we have to exclude based on other evidence so that the posterior probability of guilt following scent identification by the dog was greater than 90%? The best we can do is to reduce the pool of suspects from the original 20 to two, based on other evidence. In this case, after the dog picks one of these suspects, the posterior probability that this suspect is guilty is 88%: not quite reaching the standard for establishing guilt beyond a reasonable doubt .[7]

In practice, other evidence besides scent identification is often qualitative and not easily converted into the prior probabilities used in this example. However, if Schoon's results on the abilities of trained dogs to match human scents are accurate, the method of scent lineups doesn't seem to have much credibility. Even in the ideal and unlikely situation that other evidence reduces the pool of potential suspects to two, there is still about a 12% chance that the dog would pick the wrong suspect from a lineup.

Let's consider a parallel situation in which these kinds of calculations are helpful but actually validate a common forensic method rather than casting doubt on it. This is the well-known use of DNA to match suspects to blood or tissue samples found at a crime scene (Gomulkiewicz and Slade 1997). We can set up tables just like Tables 3.3 and 3.4 to show the calculations. With current technology, the likelihood of missing a match between a truly guilty suspect and a sample of that suspect's DNA from a crime scene is very low, certainly less than 0.5% and perhaps in principle equal to zero. This is a false negative or false mismatch, shown in the lower-left cell of Table 3.5. The probability of a false positive depends on four factors: the possibility that an innocent person shares the same DNA profile as the perpetrator of the crime for the regions of DNA that were analyzed, the possibility of laboratory error such as contamination of a sample, the possibility that the innocent person left his or her DNA at the crime scene but did not commit the crime, and the possibility that a DNA sample from the innocent person was planted at the crime scene. If the latter three possibilities can be ruled out, the probability of a false positive ranges from one in 100,000 to one in 1 billion, *for suspects who are not relatives of the perpetrator of the crime*.[8] Because brothers, for example, share 50% of their DNA, the likelihood of a false positive is as high as 0.26 for an innocent suspect who is the brother of the actual criminal. For Table 3.5, I assume that the pool of potential suspects includes only unrelated people, and I use an intermediate value for the probability of a false positive identification of one in 10 million.

Applying these values to the murder at the English country estate (and assuming that the mysterious handkerchief has blood on it that does not match that of the victim), Table 3.6 shows that the likelihood of guilt of a suspect whose DNA profile matches that of the blood on the handkerchief is

Table 3.5. Identification of suspects based on blood or tissue samples containing DNA collected at a crime scene.

		Status of Suspect	
		Guilty (Suspect = Perpetrator)	Innocent (Suspect ≠ Perpetrator)
DNA Collected at Crime Scene	Matches Suspect	0.995	0.0000001
	Does Not Match Suspect	0.005	0.9999999

The values in the table are the probabilities that DNA collected at the crime scene (1) matches that of the suspect if the suspect is guilty (the upper-left cell), (2) matches that of the suspect if the suspect is innocent (a false positive result in the upper-right cell), (3) does not match that of the suspect even though the suspect is guilty (a false negative result in the lower-left cell), and (4) does not match that of the suspect if the suspect is innocent (the lower-right cell). These values were derived from a review of the use of DNA evidence in court by Gomulkiewicz and Slade (1997).

0.995/0.9950019, which is greater than 99.99%. Even if the pool of potential suspects was much larger than 20, DNA evidence may be quite persuasive, provided factors like sloppiness in lab techniques or planting evidence at the crime scene can be excluded.

However, there may be situations in which even DNA evidence is not as conclusive as might be assumed. Suppose the *only* evidence available is DNA from a crime scene. The FBI and other law enforcement agencies have databases of DNA profiles for large numbers of individuals who have had various encounters with the legal system. The sizes of these databases are increasing daily. If the authorities have no other evidence, they may scan the database to see if there is a profile that matches that of the DNA from the crime scene. If there are 5 million profiles in the database and we assume that the guilty person is one of those 5 million, we would substitute 5 million for the overall total in the bottom right cell of Table 3.6. In this case, with one guilty person

Table 3.6. An application of DNA identification to a murder in an English country estate.

		Status of Suspect		
		Guilty (Suspect = Perpetrator)	Innocent (Suspect ≠ Perpetrator)	Totals
DNA Collected at Crime Scene	Matches Suspect	0.995	0.0000019	0.9950019
	Does Not Match Suspect	0.005	0.9999981	1.0049981
Totals		1	19	20

As in Table 3.4, we assume that there are 20 possible suspects, only one of whom is guilty. This provides the totals in the bottom row. The totals for each column are multiplied by the probabilities in the appropriate cell of Table 3.5 to get the values in this table, which can be used to compute the probability that an innocent suspect is mistakenly identified as guilty based on DNA evidence. This probability is 0.0000019/0.9950019, which is much less than 1%, if the assumptions about the use of DNA evidence discussed in the text are correct.

and 4,999,999 innocent suspects, the upper-right cell of the table would become $0.0000001 \times 4,999,999 = 0.5$. Therefore the top row of the table would be changed to 0.995, 0.5, and 1.445. Given a match between DNA from a crime scene and DNA from one of the 5 million individuals in the database, the probability of guilt for that particular individual would be 0.995/1.445, or 66%. Just as in the example of the occult-stool test for colorectal cancer, in which such a small proportion of the population has the disease that most false positive tests occur in healthy people, there is a substantial likelihood of a false positive result if DNA from a large number of innocent people is screened to find a match to DNA collected at a crime scene.

DNA evidence is certainly a powerful tool in forensic identification, but it can be misused in several ways, including the kind of fishing expedition described in the last paragraph (Roeder 1994). Although the calculations often depend on assumptions that can't be verified (e.g., that the total number of possible suspects is known), this is a valuable exercise because it introduces a systematic way of thinking about the credibility of evidence.

In this chapter we've examined several experiments dealing with the olfactory abilities of dogs in the context of forensic work. These experiments were not ideal in design and execution, but perhaps they are more valuable for learning some of the basic elements of experimentation because of their flaws than more rigorous experiments would have been. It might seem that it would be easy to study the behavior of dogs experimentally because they are so much more familiar to us than other animals, but in fact studying trained dogs means working with their handlers as well, and this can complicate experiments significantly.

Although the limited experimentation that has been done to date, mostly by Gertrud Schoon, casts doubt on the validity of scent lineups in "proving" that a suspect committed a crime, some of the other work that trained dogs do is still credible. This includes finding people buried by avalanches or collapsed buildings, sniffing out hidden narcotics, and tracking fugitives. Also, dogs clearly have the sensory ability to identify odors specific to individuals and thus determine if a suspect in a scent lineup has the same odor as an item from a crime scene, although not infallibly. The key is to devise a training method that enhances the accuracy and reliability of this identification.

Finally, this chapter illustrates a quantitative method for evaluating evidence. I used examples ranging from diagnostic tests for disease to matching DNA from a crime scene to DNA of a suspect, but this method has even broader applicability to hypothesis testing in general. In many cases, we may not have all the information necessary to apply this method. However, using this approach as a framework for thinking about the meaning of evidence can help us develop an appropriate level of skepticism about how scientific information is used in practical matters, as well as in basic research. The results of this approach are sometimes surprising, as in some of the examples we discussed. If jurors were better educated about these ideas, we would all benefit from more rational judicial proceedings.

RECOMMENDED READINGS

Gawande, A. 2001. Under suspicion: The fugitive science of criminal justice. *The New Yorker*, 8 January 2001, p. 50. Gawande briefly reviews attempts to test experimentally some basic assumptions of forensics, focusing on the veracity of eyewitness accounts of crimes.

Hoffrage, U., S. Lindsey, R. Hertwig, and G. Gigerenzer. 2000. Communicating statistical information. *Science* 290:2261–2262. This article discusses why many people have difficulty in interpreting information about probabilities, using the occult-stool test for colorectal cancer and other medical examples as illustrations.

Malakoff, D. 1999. Bayes offers a 'new' way to make sense of numbers. *Science* 286:1460–1464. The calculations in the last section of this chapter are based on Bayesian statistics, and Malakoff provides an introduction to this topic.

Taslitz, A. E. 1990. Does the cold nose know? The unscientific myth of the dog scent lineup. *Hastings Law Journal* 42:15–134. Taslitz reviews the use of dogs in legal proceedings.

Chapter 4

Why Are Frogs in Trouble?
Complementary Observations and Experiments to Test Hypotheses in Ecology

In August 1995, a group of students visited a pond in southern Minnesota on a class field trip. Tadpoles in the pond had just undergone metamorphosis to become juvenile northern leopard frogs, which were dispersing away from the pond to begin the terrestrial phase of their normal life cycle. However, a large percentage of the frogs had missing or deformed hind legs. This discovery by sixth-, seventh-, and eighth-graders and their new teacher was not the first report of deformed frogs in North America, but it sparked a great deal of public attention and led to large-scale research efforts to understand what was happening. These efforts were motivated by concerns that the deformities might have been caused by a new environmental contaminant that could also be damaging to people.

A parallel story about frogs in trouble played out in the 1990s, although it lacked the drama and intrigue of the tale of frog deformities told by William Souder (2000) in *A Plague of Frogs: The Horrifying True Story*. In 1989, scientists studying amphibians and reptiles gathered at the First World Congress of Herpetology, held in Canterbury, England. Reports by researchers studying individual populations of frogs and salamanders in various parts of the world led amphibian specialists at the conference to suspect that there might be a consistent worldwide pattern of decline and extinction of amphibian species. This hypothesis was initially challenged by some ecologists. Many amphibian populations experience wide fluctuations in population size, and most of the data showed population declines for only a few years, so critics said that apparent declines may have been only temporary. By the late 1990s, however, additional data had convinced almost all herpetologists of the reality of global declines. This story, like that of limb deformities in frogs, made the newspapers but seemed to be of more concern to biologists than to the general public, probably because the human implications were less direct than the possibility that a toxic compound in water caused deformities and might affect human health. However, population declines and extinction of amphibians could be an early warning sign of human impacts on natural environments without indicating a specific health problem for humans (Blaustein and Wake 1995).

"Muddy-boots" biologists vigorously tackled the problem of characterizing these two phenomena. Basic questions about the prevalence of deformities in frogs and the status of amphibian populations in general could only be answered by field biologists willing to tromp around in ponds and swamps year after year. Although biologists with these propensities are relatively fewer than in the heyday of natural history a century ago, their contributions are essential for solving problems like these. As descriptive data accumulated, several hypotheses about the causes of deformities and of declining populations emerged. For each phenomenon, the main alternative hypotheses had markedly different implications for humans. The two major hypotheses for frog deformities were that they were caused by parasites or by a toxic chemical in the environment. If parasites were the culprit, there would be no reason for direct concern about human health because the kinds of parasites that can cause limb deformities in frogs are not transmitted to humans, although parasites might be a symptom of wetland degradation that could have adverse long-term consequences. However, toxic chemicals in wetlands are also likely to be in water supplies used by people. Chemicals that cause abnormal development of frogs could affect reproduction and development of humans because the basic biochemical processes are similar in all vertebrates.

In the case of population declines, one general hypothesis was that there is no common cause for all populations of amphibians that are threatened or endangered. In one area, habitat destruction might be the critical factor; in another, an introduced species of fish might be preying on frog larvae; in yet another, pollution might be the key. If societies and governments around the world can be convinced that conserving amphibians and other species is worthwhile, this hypothesis implies that different kinds of action will be required in different areas. The second general hypothesis was that global climate change is a common underlying cause of most recent declines and extinctions of amphibian populations. This hypothesis doesn't minimize the importance of a broad-scale approach to conservation: protecting habitats, controlling introduced species, and reducing pollution. But if true, it emphasizes the additional importance of limiting emissions of greenhouse gases that contribute to global climate change. Advocates of this hypothesis suggest that the worldwide decline of amphibian populations that has been documented in the past 20 years is an early symptom of pervasive effects of global climate change on natural environments.

These two stories about the troubled lives of amphibians illustrate the use of complementary kinds of evidence to evaluate hypotheses. In particular, the combination of key experiments with critical natural observations can be an especially powerful approach in science. In Chapter 2, I discussed both non-experimental and experimental evidence relating to the hypothesis that vitamin C has various benefits for human health. But this example wasn't as suitable as the examples to be discussed in this chapter for showing how researchers *integrate* observational and experimental evidence in testing hypotheses. In addition, the amphibian examples illustrate some of the rewards of fieldwork in biology.

WHAT CAUSES LIMB DEFORMITIES IN FROGS?

The 1995 discovery of deformed northern leopard frogs by Minnesota school-children inspired a large number of researchers to start searching for the cause of this phenomenon. Some of these researchers had worked on amphibians for years; others brought different kinds of expertise to the problem. This group generated several hypotheses, but two seemed the most credible—that deformities were caused by a parasitic flatworm called a trematode or by a type of chemical called retinoic acid that is related to vitamin A. Attention quickly focused on these hypotheses for two major reasons. First, previous work had shown that trematodes were associated with deformities in a population of frogs in California, and high doses of retinoic acid were known to be powerful teratogens in humans and other animals (teratogens are compounds that cause abnormal development).[1] This background information gave these two hypotheses a higher level of plausibility than some of the others, which in turn encouraged researchers to devote more effort to testing these hypotheses. Second, the parasite and retinoic acid hypotheses were specific enough that researchers could figure out reasonable ways to test them.[2]

Most work on retinoic acid had been done by developmental biologists in the laboratory. Workers such as Susan Bryant and David Gardiner at the University of California at Irvine used amphibians as a model system to study the development of limbs in vertebrates. One powerful way to study what happens in *normal* development is to use chemical or mechanical treatments that produce *abnormal* development, and Bryant and Gardiner had pioneered this approach by using retinoic acid and many other means to alter limb development in amphibians (Souder 2000). But credibility of the retinoic acid hypothesis for the occurrence of limb deformities in natural populations of frogs depends on identifying a possible source of increased concentrations of retinoic acid in the wetlands where deformed frogs are found. Researchers quickly settled on a widely used insecticide called methoprene, which prevents metamorphosis of larval insects into adults. Methoprene was licensed for use in the United States in 1975 as a substitute for DDT and other chlorinated hydrocarbons that were banned a few years earlier. Scientists assumed that methoprene was much safer than chlorinated hydrocarbons because it breaks down rapidly and because it acts by mimicking a hormone that is specific to insects. But frogs and other vertebrates aren't immune to its effects because one of the breakdown products of methoprene is a compound that interacts with molecular receptors for retinoic acid in the membranes of vertebrate cells. And, indeed, high doses of methoprene cause limb abnormalities in mice.

This effect makes the retinoic acid hypothesis a strong candidate to explain the increased frequency of limb deformities that began to be seen in natural populations of frogs in the 1990s. However, G. T. Ankley and several other scientists with the United States Environmental Protection Agency (EPA) did a laboratory experiment with northern leopard frogs that cast doubt on the hypothesis. They exposed batches of eggs to various concentrations of methoprene in the presence and absence of ultraviolet light, which might induce the breakdown of methoprene into a form that functions like retinoic acid. Although high doses of methoprene caused severe developmental abnormalities and death of all tadpoles, methoprene did not produce hind limb abnormalities specifically (Ankley et al. 1998).

What about the parasite hypothesis? How could trematodes possibly cause limb abnormalities in frogs? Many parasites have complex and fascinating life cycles, with various stages that infect different kinds of hosts. Some kinds of parasites infect humans, with devastating consequences for individuals and societies. For example, about 200 million people in tropical areas around the world are afflicted with schistosomiasis, a disease that can produce permanent damage to the bladder, liver, other intestinal organs, and sometimes the brain. Schistosomiasis is caused by a trematode that infects people who bathe or wade in water containing the parasite, for which snails are the intermediate host. Parasite eggs released in the urine or feces of humans are picked up by snails, where they develop through several stages and eventually form larvae that are released into the water. The larvae burrow through the skin of humans, where they migrate to blood vessels in the abdominal organs and become adults. The adults release eggs, completing the cycle.

Trematodes that infect frogs have a similar life cycle, again with snails as an intermediate host and, in some cases, fish as an alternative vertebrate host (unlike schistosomiasis, these parasites don't infect people). Most frogs lay their eggs in water, and the eggs hatch into tadpoles. The tadpoles remain in the water to feed and grow, and they eventually metamorphose into frogs, which leave the water to live on land. Larvae of trematode parasites infect tadpoles by burrowing under the skin, and some species have a particular propensity to do so near the base of the hind limbs. When a clump of these larvae collects near the bud of a developing limb, this can disrupt normal development of the limb by interfering with the information transmitted between cells in the developmental process. In fact, Stanley Sessions and S. B. Ruth (1990) simulated the production of abnormal hind legs in frogs and salamanders by inserting glass beads in the limb buds in the same positions at which they found concentrations of trematode larvae. In short, when scientists began to think about possible causes of high frequencies of abnormalities in frogs, the parasite hypothesis had a good deal of plausibility.

Pieter Johnson and three colleagues (1999) thoroughly tested the parasite hypothesis by combining critical experiments in the laboratory with trenchant observations in the field. Johnson was an undergraduate at Stanford University when he began his studies of limb deformities in Pacific treefrogs. His group found Pacific treefrogs at 13 ponds near Stanford. Deformed frogs were present at high frequencies in four ponds; moreover, an aquatic snail that is the intermediate host for a trematode parasite called *Ribeiroia* was present in the same four ponds, and only in these ponds. Johnson's group also found larvae of the trematodes in frogs in all four ponds.

The researchers didn't directly contrast the parasite hypothesis and the retinoic acid hypothesis. However, they tested the water from two of the ponds with deformed frogs for pesticides, PCBs, and heavy metals, and found none. More important, because no chemical assay of water can test for all possible contaminants, Johnson's group collected 200 eggs of Pacific treefrogs and brought them into the lab with water from the same ponds. These eggs hatched, and the resulting tadpoles metamorphosed into frogs without limb abnormalities.

These field data implicated a particular trematode parasite, *Ribeiroia*, as a cause of limb deformities in Pacific treefrogs, but Johnson's group wanted to test this hypothesis more rigorously. Therefore they designed an experiment to infect tadpoles with *Ribeiroia* in the laboratory. They collected treefrog eggs from a river in northern California where there had been no reports of abnormalities. When the eggs hatched, the resulting tadpoles were kept individually in aquaria containing commercial spring water, which presumably was free of parasites and pesticides. Johnson's group used four basic treatments: no exposure to *Ribeiroia* larvae (a control treatment), light exposure (16 parasites), moderate exposure (32 parasites), and heavy exposure (48 parasites). These levels represented the range in numbers of larvae that they had found infecting frogs in nature, contributing to the realism of their laboratory experiment.

Johnson and his colleagues found no deformities among the 31 frogs that developed from tadpoles in the control treatment, but they did find deformities in a majority of frogs exposed to 16 parasites as tadpoles and virtually all frogs exposed to 32 or 48 parasites. Furthermore, the types of deformities were similar to those seen in natural populations of Pacific treefrogs in ponds with *Ribeiroia*. In natural populations, 95% of deformities involved the hind limbs, and 53% of these involved extra legs. In experimental treatments, all observed deformities involved the hind limbs. Extra legs accounted for 32% of deformities in the light-exposure treatment, 44% in the moderate-exposure treatment, and 55% in the heavy-exposure treatment. Two details of these results should increase our confidence that this experiment was a realistic simulation of what occurs in nature: the fact that the range of deformities seen in the experiment was similar to that seen in frogs in the field, and the fact that the likelihood of developing extra hind legs increased with the degree of exposure to parasite larvae in the lab.

Johnson's group used two other treatments in their experiment that were particularly revealing. In one, they exposed tadpoles to another species of trematode called *Alaria;* in the second, they exposed tadpoles to both *Ribeiroia* and *Alaria*. What was the purpose of these treatments? In their field sampling,

Johnson's group had found other parasites, including *Alaria*, that were infecting frogs, but *Ribeiroia* was the only parasite concentrated around the base of the hind legs. Therefore, if *Ribeiroia* was the specific cause of limb deformities, the researchers predicted that infection of tadpoles by *Alaria* in the lab wouldn't cause deformities. On the other hand, if deformities were simply a general response to stress that could be induced by various factors, *Alaria* should cause deformities in the lab. In fact, even though the researchers used almost twice as many larvae of *Alaria* as in the heavy-exposure treatment with *Ribeiroia*, there were no limb deformities in frogs exposed as tadpoles to *Alaria* alone and there was a similar number in frogs exposed to both parasites as in frogs exposed to *Ribeiroia* alone. In the lab experiments, as in the field, *Alaria* infected tadpoles but were not concentrated in the pelvic region.

This study by a group of young biologists is a nice illustration of integrating field observations and laboratory experiments to test an hypothesis. A consistent and coherent story about relationships among Pacific treefrogs, parasitic trematodes, and snails as alternative hosts of the trematodes emerges from the research by Johnson's group. The study leaves some questions unanswered, however. One of the most intriguing is, Why do *Ribeiroia* larvae burrow into the pelvic region of tadpoles to become concentrated near the hind limb buds, when other trematode parasites do not have this behavior? Stanley Sessions and S. B. Ruth (1990) suggested that this might be an evolutionary adaptation of the parasite to increase its likelihood of transmission to other vertebrate hosts. If the parasite causes frogs to develop missing or deformed hind legs, this should make these frogs more vulnerable to predators, which would themselves become infected after eating the frogs. This hypothesis has not been thoroughly tested.

A second question left unanswered by Johnson's study is, How much can we generalize from these results on Pacific treefrogs in California? After all, the initial concern about limb deformities in frogs came from observations of different species in Minnesota and other places far removed from California. It's possible that retinoic acid or some other chemical causes deformities in various species of frogs in the former sites, whereas parasites cause deformities in Pacific treefrogs in California. Unfortunately, it's more difficult to come up with universal solutions to problems in ecology than in molecular biology. However, Pieter Johnson and nine other researchers (2002) recently described results of a survey of 101 sites in five western states in which they examined about 12,000 amphibians of 11 species for various kinds of deformities. *Ribeiroia* was always found at sites where one or more species of amphibians had a high frequency of deformities, but *Ribeiroia* was usually absent where frequencies of deformities were low in all species. The snails that are the intermediate hosts of *Ribeiroia* were also more abundant at sites where amphibians were infected with the trematodes. Johnson's group also measured the concentrations of 61 different herbicides and insecticides at their sites. Only a few sites had measurable concentrations of any of these pesticides, and there were no relationships between pesticide concentrations and abnormalities in amphibians.

A third interesting set of questions raised by Johnson's study is this: If parasites cause deformities in Pacific treefrogs or other species, are there environmental factors that have changed to cause an increase in the abundance of parasites or in the sensitivity of frogs, so that the frequency of limb deformities in natural populations has increased dramatically in recent years? If so, what are these environmental factors? Joseph Kiesecker (2002) of Pennsylvania State University provided a possible answer to these questions in a study that was especially persuasive because it linked field and laboratory experiments, this time with wood frogs in Pennsylvania. Kiesecker did his field experiment at six ponds, three of which contained the pesticides atrazine and malathion, due to agricultural runoff, and three of which did not contain measurable amounts of these pesticides. He placed young tadpoles in mesh cages at each pond. Some of these cages were covered with fine mesh that prevented trematode parasites from entering and infecting the tadpoles, and some cages had coarser mesh that did not prevent the parasites from entering. Tadpoles developed limb deformities in all cages that allowed parasite access, but the frequency of deformities was substantially greater in the ponds that received agricultural runoff. In a complementary lab experiment, he exposed tadpoles to pesticides plus trematode parasites, pesticides alone, parasites alone, or neither pesticides nor parasites. The highest frequency of deformities occurred in the treatment with both parasites and pesticides, and analysis of blood samples suggested that the pesticides might weaken the immune system of the tadpoles, making them more vulnerable to the parasites that actually cause deformities.

The most sobering aspect of Kiesecker's results is that atrazine and malathion contributed to hind limb deformities in frogs at concentrations low enough to be classified as safe for drinking water by the Environmental Protection Agency (EPA). Although the direct cause of these deformities was a parasite that doesn't infect humans, it may be that very low concentrations of these pesticides compromise our own immune systems, as suggested by Kiesecker for wood frogs. Another recent study, by Tyrone Hayes of the University of California at Berkeley, implicates atrazine independently of parasites in a different kind of abnormality in frogs. Hayes and his colleagues (2002) found that 36% of male leopard frogs exposed as tadpoles to 0.1 parts per billion of atrazine in the laboratory developed abnormal testes. Many of these males were feminized as adults; that is, they produced eggs, as well as sperm. Atrazine is the most widely used herbicide in the United States, and the EPA allows 30 times as much atrazine in drinking water as Hayes's group used in their experiment. Other research has shown that atrazine can induce synthesis of female sex hormones in male fish, reptiles, and mammals, so these new data on frogs have raised alarms, although other researchers have not yet been able to replicate the effects of very low levels of atrazine found by Hayes and his colleagues.

In this section I have focused on studies of the parasite hypothesis because they integrated field observations and laboratory experiments. This is a powerful approach because it combines the rigorous elements of a well-designed experiment with the realism of studying phenomena under natural conditions. Although the parasite hypothesis initially seemed to have fewer implications for human health than the alternative hypothesis, that deformities were caused by toxic pollutants in wetlands, the recent research by Kiesecker suggests that two specific pesticides might interact with parasites to cause frog deformities. The full implications for human and environmental health of this epidemic in North America are still unclear, but many researchers continue to study the problem and there will undoubtedly be substantial progress in understanding the ramifications of developmental abnormalities in wild populations of amphibians in the near future.

WHAT CAUSES AMPHIBIAN POPULATIONS TO DECLINE?

Much work has also been done on the more comprehensive issue of worldwide declines in amphibian populations. To review, two general hypotheses for these declines could be considered umbrella hypotheses because each can take many specific testable forms. The first general hypothesis is that each case of a population that has become rare or extinct has its own set of causes and should be investigated independently of other cases. The second is that these worldwide declines of amphibian populations have a common underlying cause in some aspect of global environmental change. According to the first hypothesis, many specific factors can affect amphibian populations: habitat destruction, pollution, introduced species that prey on native amphibians or compete with them, and so on. Although declining populations are widespread, this doesn't implicate a common cause, but rather illustrates the many different ways in which humans can affect natural environments to the detriment of amphibians and other species. Advocates of the second hypothesis don't deny that all of these factors influence amphibian populations; in fact, almost everyone sees habitat destruction as the primary reason that many amphibian populations are in jeopardy. However, these advocates also suggest that some aspect of environmental change that has global scope exacerbates the danger.

One important consequence of these different views is practical. If causes are all localized, solutions must be localized as well. Conservationists must work on many fronts, in many different ways, to preserve amphibians. However, if there is a common global cause for declines, such as global warming, solutions will require broadly based global action *in addition to local initiatives*. This is a more challenging problem. Beyond this practical implication is an additional philosophical issue. There is tension within ecology in general between a viewpoint that seeks broad generalizations to explain fundamental phenomena and one that emphasizes the uniqueness of each species in every environment. Are there generalizations to be found in ecology that are comparable in scope to Einstein's theory of relativity or Darwin's theory of evolution by natural selection? Or must ecologists be satisfied with collecting case studies from which no generalizations emerge?

There are about 4,000 species of amphibians (frogs, toads, and salamanders), but only a small fraction has been studied in enough detail to determine if their populations are declining. There certainly are dramatic examples of precipitous declines and extinctions of some species. One classic case is the golden toad, a stunningly beautiful species that was discovered in the highlands of Costa Rica in 1964. Golden toads occupied only a few square miles of cloud-forest habitat in the Monteverde Reserve. They were abundant in this small area as recently as 1987 but declined from 1,500 animals in that year to only one in 1988 and 1989. No golden toads have been seen since 1989, despite the fact that they lived in a nature reserve protected from overt disturbance and intensively studied by tropical ecologists. Another example is the gastric-brooding frog in Australia. Females of this species swallow their own eggs, then fast for six weeks while the eggs hatch and develop into baby frogs in their stomachs. This species was discovered in 1973, was reported to be abundant in the Cannondale Ranges of Queensland in 1976, and had disappeared by 1980 (Sarkar 1996).

Are these representative examples or unusual cases? A group of researchers led by Jeff Houlahan (2000) of the University of Ottawa scoured the literature and surveyed herpetologists to find as much data as possible on changes in sizes of amphibian populations over time. These are called time-series data, and they come from field biologists who visit the same sites year after year and estimate the numbers of animals present.[3] Houlahan's group found such data for 936 populations of 157 different species of amphibians that had been collected by more than 200 researchers. When they published their analyses of these data in 2000, this was the most complete set of time-series data for amphibians that anyone had examined.

Most of these time series were relatively short. The average was seven years and the longest was 31 years. Nevertheless, the researchers reported that most populations were declining and that 61 of them had become extinct while they were being studied (extinction of a population does not necessarily imply extinction of the species to which it belongs because most species are represented by many different populations). Amphibian populations were declining in all regions of the world, although almost 90% of the time series available to Houlahan's group were from western Europe and North America. Another indication of increasing threats to amphibian populations is that 87 North American species were classified as threatened or endangered in 1998, compared to 38 in 1980.

If there is a common characteristic of amphibian populations that are losing ground, it is that they tend to occur at relatively high elevations. Much of the detailed research that has gone beyond simply documenting population declines to try to understand their causes has been done in the Cascade and Sierra Nevada mountain ranges of the western United States, although recent work in the highlands of Central and South America and Australia has broadened understanding of the problem. To illustrate the integration of experimental and observational approaches in ecology, I'll describe a set of studies by Andrew Blaustein and his students and colleagues in the Cascade Range of central Oregon. Like the research by Johnson on limb deformities in Pacific treefrogs, Blaustein's work shows the productive synergy that can exist between field observations and critical experiments. In the course of long-term field studies on several species of amphibians, which he began in the late 1970s, Blaustein developed a series of hypotheses for population declines that he tested with carefully and cleverly designed field experiments.

Blaustein and his colleagues (1994a) focused on three species that breed in lakes at relatively high elevations in the Cascades: Pacific treefrogs, Cascades frogs, and western toads (Figure 4.1). After mating, females of all three species lay their eggs in shallow water at the edges of lakes. Western conifer forests are not particularly dense, so the eggs get plenty of sunlight as they hatch and begin to develop into tadpoles. Historical records beginning in the 1950s indicate that at least 90% of the eggs of each species typically survived through metamorphosis each year. This continued through the early phase of the intensive studies by Blaustein's group, but in the mid-1980s survival of the eggs of Cascades frogs and western toads declined dramatically to less than 50%. Egg survival has remained low through the present for these two species but stayed as high as it was in the early years for Pacific treefrogs.

Frogs and toads are exposed to mortality from many different sources during the various stages of their life cycle. A female may lay hundreds or thousands of eggs each year during her lifetime, only two of which have to survive to reproductive maturity on average to maintain a stable population. From this perspective, it may not matter whether 10% or 90% of eggs survive through metamorphosis; if mortality of adults is low enough, a population could be stable or grow rapidly in size even with high egg mortality. Therefore, we can't just assume that the dramatic decrease in survival of eggs after the mid-1980s translated into population declines for Cascades frogs and western toads; we need to examine this directly by looking at population data for adults. In fact, both species did decline, to the point that they became candidates for listing as threatened species by the U.S. Fish and Wildlife Service. Pacific treefrogs, on the other hand, maintained healthy populations through this same period.

These contrasting patterns of egg mortality and population dynamics in three species of amphibians illustrate why ecology is such a rich source of questions for scientific study. What differences in the biology or environments of the species could account for the fact that Pacific treefrogs are doing well but the other two species are in trouble? There are many possibilities, but the initial observation of a difference leads to a powerful strategy for attacking the problem. Imagine that herpetologists observed an increase in egg mortality in all species of amphibians. It would be difficult to know where to start in searching for a cause of this phenomenon. However, if egg mortality increases in some species but not in others, our search for a cause can be narrowed down to factors that differ among the species. For example, if the species laid eggs in different types of habitats, some feature of the habitat would be a strong candidate for the cause of differential mortality. In fact, Pacific

Figure 4.1. Pacific treefrog (*Hyla regilla*, top), Cascades frog (*Rana cascadae*, middle), and western toad (*Bufo boreas*, bottom). Photographs taken by Grant Hokit and used with his permission.

treefrogs, Cascades frogs, and western toads occur together in some lakes and all three species tend to lay eggs in shallow water at the edges of these lakes, yet great differences in mortality of eggs occur. This suggests that some aspect of the biology of the species, rather than of the environments where they occur, accounts for the differences in mortality.

One of the first things Blaustein did when he began to notice increased mortality of eggs of Cascades frogs and western toads was to bring eggs and lake water to his laboratory at Oregon State University to see how successfully the eggs hatched and developed into tadpoles. If there was a new pollutant or some other chemical change in the water that was particularly detrimental to these two species, it could explain the increased mortality of eggs after 1985. Furthermore, this mechanism would be reflected in reduced survival of eggs in the laboratory also, as long as the eggs there were kept in natural lake water. But Blaustein and his colleagues (1998) got greater than 99% survival in this simple lab experiment, implying that polluted lake water wasn't the problem.

If bad water wasn't causing egg mortality, what differences between the natural environment and the lab environment might be candidates? Fairly early in the process of studying this problem, Blaustein hypothesized that increased exposure to ultraviolet radiation in the 1980s might be the culprit. A specific type of ultraviolet radiation called UV-B can be very damaging to cells and tissues; this is what causes sunburn in humans, for example. The UV-B kills or damages cells by altering the chemical structure of DNA and proteins. The ozone layer in the upper atmosphere absorbs much of the UV-B in sunlight, but a variable proportion of this radiation reaches Earth's surface, depending on season, time of day, cloud cover, and other factors. Release of chlorinated fluorocarbons, which used to be used as cooling agents in refrigerators, has depleted the ozone layer. The most dramatic example is the ozone hole that develops over Antarctica each year, but ozone has also been depleted at temperate latitudes over the northern hemisphere (Christie 2000).

In developing the UV-sensitivity hypothesis for increased mortality of amphibian eggs, Blaustein and his colleagues (1995) reasoned that these eggs were relatively transparent, meaning that they would absorb more ultraviolet radiation than if they were opaque; they were often laid in shallow water in open areas subject to full sunlight for many hours of the day; and egg mortality typically occurred at relatively high elevations, where the intensity of solar radiation is high. The fact that eggs developed normally in the laboratory was consistent with the UV-sensitivity hypothesis, but it was not conclusive because many other factors besides exposure to ultraviolet radiation would differ between the lab bench and the outdoors. Blaustein decided that a key test would be to alter the exposure of eggs to UV, so he designed a field experiment to do this. The plan was to expose some sets of eggs to normal conditions in the field, including normal levels of UV radiation, and protect other sets of eggs from UV.

Blaustein's group (1998) has done this experiment at numerous sites with at least eight species of amphibians, including frogs, toads, and salamanders, but I'll focus on the work with Pacific treefrogs, Cascades frogs, and western toads. The UV-sensitivity hypothesis leads to two basic predictions. If it is true, protecting eggs of Cascades frogs and western toads from ultraviolet radiation should increase survival rates. However, this treatment shouldn't make a difference for Pacific treefrogs because their survival rates are high even when exposed to natural levels of UV.

The experimental design was straightforward, although the logistics of conducting the experiment at remote sites in harsh weather during early spring were challenging (Blaustein, 1994, gave a personal account of the pleasures and perils of fieldwork in an article published in *Natural History*). Blaustein's group built a large number of enclosures in which to place eggs in the field. These were boxes made of Plexiglass for the sides and fine-mesh screening on the bottom. Each enclosure was about 15 inches square and almost 3 inches tall. The researchers used three types of enclosures to implement three treatments. They used boxes with no tops for a control treatment; eggs in these boxes would be exposed to normal levels of sunlight. For the treatment with reduced UV exposure, they used boxes with mylar filters on top, which blocked 100% of the incident UV-B radiation. Finally, they used boxes with acetate filters on top as a third treatment. These filters only blocked about 20% of the UV radiation; this treatment was used as a control for any effect that covers might have on the development of eggs separate from the ability of mylar filters to block UV.

The experiment consisted of placing 150 eggs of a single species in an enclosure and monitoring these eggs until all of them had either hatched into tadpoles or died. Blaustein and his colleagues (1994a) did the experiment at four lakes in central Oregon. All three species were studied at Three Creeks Lake, and each species was studied individually at one other lake. For each species at each site, they used 12 enclosures, four for each treatment. In other words, they used two kinds of *replication* in their research. Each treatment was replicated four times for each species at each site, and each species was studied at two sites. Replication of treatments at one study site is important as a way of assessing the effect of random variation in other environmental factors compared to variation in the manipulated factor. Suppose you used only three enclosures for western toads at Three Creeks Lake and found that egg survival was higher in the enclosure with the mylar filter than in the two control enclosures. This might be due to the blocking of UV radiation by this filter, as predicted by the UV-sensitivity hypothesis. But it might also be due to the fact that, by chance, this particular enclosure had been placed in a more favorable location along the shore of the pond than the other two enclosures. With only one enclosure with UV blocked and one of each type of control, there would be no way to know for sure. But if the treatments are replicated and randomly interspersed at the study site, the various replicates of each treatment should experience a range of the environmental conditions that characterize the site. By comparing the variation in egg survival *between treatments* to the variation *among replicates within each type of treatment*, Blaustein's group could measure the effect of the treatment relative to the effects of other environmental factors that might affect egg survival. This in turn enabled them to assess the importance of UV exposure for survival of the eggs.

This kind of replication of treatments at a single site is common in ecological field experiments. Replication of an experiment at different sites is less common but also valuable as a way of testing the general validity of the experimental results. If Blaustein's group had done an experiment with one species at a single site and got results consistent with the UV-sensitivity hypothesis, a skeptic might argue that there was something peculiar about that site that made eggs of the species studied particularly sensitive to UV. If results are unique to one species at one site, they are less interesting than if they are found more widely.

The results of this experiment were unambiguous. Protection from UV radiation increased the survival of western toad eggs from less than 60% to about 85% at both sites where they were studied. There were similar improvements in survival of Cascades frog eggs, but eggs of Pacific treefrogs had survival rates greater than 95% in control enclosures, as well as in enclosures protected from UV. These results are consistent with the differences in survival under natural conditions that Blaustein's group had observed in the years before they did their experiment. Evidently, something about the biology of Pacific treefrogs protects their eggs from UV damage, whereas similar amounts of ultraviolet radiation cause eggs of the other two species to die. What difference between the species could account for their differential sensitivity to UV?

Blaustein collaborated with molecular biologist John Hays to answer this question. Organisms have mechanisms for repairing damage to the DNA in their cells. One of these mechanisms involves an enzyme called photolyase, and Hays had been studying how photolyase works in another amphibian, the African clawed frog, that is used widely in laboratory studies. Having developed techniques to measure photolyase in eggs of this species, it was straightforward to measure the enzyme in eggs of the three species Blaustein was studying in the field. Hays found that photolyase activity in eggs of Pacific treefrogs was six times as great as in eggs of western toads and three times as great as in eggs of Cascades frogs (Blaustein et al. 1994a). It's not clear why there was this difference between the three species since all of them lay eggs in shallow water with high exposure to solar radiation. Nevertheless, the difference is congruent with the observed differences in egg survival under natural conditions and with the results of the field experiment done by Blaustein and his colleagues.

This research was criticized on various grounds. Some of these criticisms reflected misunderstanding of the experimental design. For example, the Canadian herpetologist Lawrence Licht (1996) wrote a letter to *Bioscience* entitled "Amphibian Decline Still a Puzzle" in response to a rebuttal by Blaustein and his colleagues (1995) of an earlier critique by Licht (1995). In his letter, Licht stated, "I also understand an attempt was made to control all variables but UV-B transmission, yet unfortunately, temperature was not controlled and remains a major problem" (Licht 1996:172). This complaint misses the point of the experiment and overlooks a fundamental strength of field experiments in general. In fact, the only variable that *was* controlled by the use of different kinds of filters on the tops of the enclosures was UV-B transmission. Other environmental variables such as temperature varied naturally among enclosures. The real question asked by the field experiment is, Can egg survival be improved by blocking UV radiation, even when other factors vary within the normal ranges that occur in the natural environments in which eggs are laid? If so, this is strong support for the UV-sensitivity hypothesis because the effect of blocking UV is apparent not only in very specific, controlled conditions used in a laboratory experiment but also in the variable conditions that typically occur in nature.

The most significant criticism of the work by Blaustein's group was that they didn't have measurements of ultraviolet radiation at their study sites to show that increased mortality of eggs in the mid-1980s corresponded to increases in amounts of UV radiation at that time. However, Blaustein didn't think of the UV-sensitivity hypothesis until after egg mortality started to increase, and it was impossible to go back in time and measure UV before this happened. In addition, it would have been difficult to collect data on the precise amount of UV to which eggs were exposed because this can vary minute by minute as clouds pass overhead, a tree branch swaying in the wind temporarily shades a clutch of eggs, or some other event alters the amount of sunlight hitting the water. In any case, the lack of long-term data on levels of UV radiation in the Oregon Cascades does not invalidate the key results of Blaustein's experiments in which exposure of eggs to UV was manipulated in enclosures by using protective filters.

The field experiment by Blaustein's group suggests that ultraviolet radiation contributes to egg mortality in some species of amphibians, although it doesn't conclusively show that increased UV radiation in the mid-1980s caused the dramatic increase in egg mortality in natural populations that the researchers observed at that time. It also doesn't explain the mechanism by which ultraviolet radiation influences the viability of eggs. It's possible that UV kills eggs directly, by damaging proteins and DNA in the cells of embryos as they develop. Alternatively, it may be that UV acts indirectly to increase egg mortality by making the eggs more susceptible to another mortality factor. For example, UV-B radiation has detrimental effects on immune responses of vertebrates, so perhaps it contributes to amphibian mortality by making developing embryos more sensitive to pathogens.

In fact, at about the time the researchers were doing their field experiment, they also documented the impact of the fungus *Saprolegnia ferax* on eggs of western toads at Lost Lake (Blaustein et al. 1994b). This fungus is a major pathogen of fish throughout the world but had rarely been found in amphibians. Blaustein's group monitored breeding by western toads at Lost Lake in 1992. There were 208 females in the population that year, and all laid their eggs in two huge masses in an area of about 15 × 30 feet between April 26 and April 29. This concentrated breeding and egg laying is characteristic of western toads. Blaustein's group estimated that about 2.5 million eggs were laid, and they first noticed fungus on some of the eggs on April 28. One week later, at least 70% of the eggs were dead, and the ultimate mortality rate was 95%.

These observations, together with their experimental work raising eggs in enclosures with differential exposure to ultraviolet radiation, led Blaustein's group to design a new experiment to compare the effects of UV radiation and *Saprolegnia* on mortality of amphibian eggs. Since the initial experiment showed that UV exposure could increase egg mortality and the observations at Lost Lake in 1992 strongly implicated *Saprolegnia*, the researchers were particularly interested in whether these two factors had a *synergistic* effect on egg mortality, with UV radiation increasing the sensitivity of the eggs to the pathogenic fungus.

Kiesecker and Blaustein tested this hypothesis in an experiment reported in 1995. The design paralleled that of the test of the UV-sensitivity hypothesis alone that we have already considered: they tested three species of amphibians, each at two sites, and they used three types of enclosures, one with a UV-blocking filter, one control with an open top, and one control with a clear top that transmitted UV radiation. But they did these experiments in plastic wading pools containing natural lake water that were placed on the adjacent shore. Half of the pools were inoculated with a culture of *Saprolegnia* fungus; half were treated with an antifungal compound to destroy any *Saprolegnia* that might have been naturally present. They used a total of 24 pools for each species at each site: four containing enclosures with UV-blocking filters and inoculated with fungus, four containing enclosures with UV-blocking filters and the antifungal compound, four containing open enclosures and inoculated with fungus, and so on. This is called a *factorial experiment* because they were testing all combinations of two factors. The first factor was the presence or absence of *Saprolegnia*; the second factor was exposure to UV. This is a standard type of experimental design for testing whether different factors have independent or synergistic effects on a process of interest.

For western toads at Lost Lake, Figure 4.2 shows that the presence of the fungus reduced egg survival from about 95% to 85% when the eggs were not exposed to UV, but from 95% to 50% when eggs were exposed to UV. Ultraviolet exposure alone apparently didn't cause decreased survival of western toad eggs at this site; exposure to the fungus alone caused a moderate decrease in survival under conditions of the experiment, and this negative effect of the fungus was greatly exacerbated by UV light. These results are consistent with the synergistic hypothesis. Kiesecker and Blaustein found a similar pattern for western toads at Three Creeks Lake and for Cascades frogs at two sites. For Pacific treefrogs, however, only the fungus affected the survival of eggs, as Kiesecker and Blaustein had predicted based on the higher concentration of photolyase, which protects against UV damage in Pacific treefrogs (Figure 4.2).

The story has gotten more complex and more interesting with this evidence that UV radiation and a pathogenic fungus interact to produce a synergistic effect on survival of eggs of western toads and Cascades frogs. But the question of what changed in the 1980s to result in decreased egg survival and declining populations still hasn't been resolved. It might be that the fungus has always been present, but exposure to UV radiation increased because of the thinning of the ozone layer in the atmosphere or some other factor. Or it could be that there was no significant change in exposure to UV radiation, but the fungus was introduced to places like Lost Lake when hatchery-reared fish were released in the lake. Or both of these things may have occurred. Testing these possibilities required looking at historical observational data because Blaustein's group was trying to explain something that had already occurred.[4]

Kiesecker and his colleagues presented some revealing data to help resolve this issue in an article published in the journal *Nature* in April 2001. They

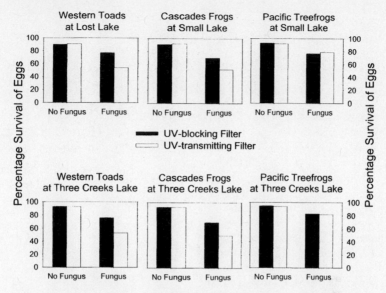

Figure 4.2. Effects of a pathogenic fungus, *Saprolegnia*, and exposure to ultraviolet radiation on average survival of eggs of three species of amphibians at three lakes in the Cascade Range of central Oregon. The filled bars show results for enclosures covered by UV-blocking filters; the open bars show results for control enclosures covered by UV-transmitting filters. There was another set of control enclosures with no covers; the results for these enclosures were similar to those for enclosures with filters that transmitted ultraviolet radiation. Modified with permission from Figure 1 of "Synergism between UV-B Radiation and a Pathogen Modifies Amphibian Embryo Mortality in Nature," by J. M. Kiesecker and A. R. Blaustein, *Proceedings of the National Academy of Sciences USA* 92:11,049–11,052, copyright ©1995 by the National Academy of Sciences.

studied western toads from 1990 through 1999 at several sites in the Cascade Range, finding much better survival of eggs laid in deep water than in shallow water (Figure 4.3A). Virtually all mortality of the eggs was attributed to infection with fungus. Survival was probably greater in deep water because UV radiation doesn't penetrate very deeply in water, whereas UV exposure in shallow water decreases the resistance of eggs to infection by the fungus. Indeed, in an experiment complementing this primarily observational study, Kiesecker and his colleagues placed eggs in enclosures covered by UV-blocking filters and in control enclosures at three depths in Lost Lake. They found high and equivalent survival in the two types of enclosures at depths of 50 centimeters and 100 centimeters but much lower survival in the control enclosures than in the enclosures with a UV-blocking filter at 10 centimeters. They also found that the intensity of UV-B radiation dropped off steadily from the surface of the water to a depth of 100 centimeters.

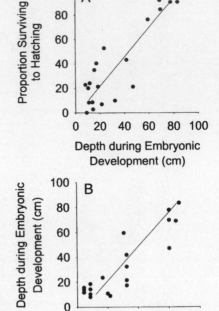

Figure 4.3. Effects of environmental conditions on the reproduction of western
toads in the Cascade Range of Oregon.
Panel A shows the relationship between
the water depth at which western toad
eggs were laid in various lakes and the
proportion of eggs that survived to hatch
because they were not killed by the fungus *Saprolegnia*. Panel B shows the relationship between water depth in these
lakes and winter precipitation in the Cascades. Panel C shows the relationship
between an index of the severity of El
Niño conditions, SOI (the Southern
Oscillation Index), for the summers of
1990–1999 and precipitation the following winter. There are more points in A
and B than C because several lakes were
sampled in most years. Modified with
permission from "Complex Causes of
Amphibian Population Declines," by
J. M. Kiesecker, A. R. Blaustein, and
L. K. Belden, *Nature* 410:681–684, copyright ©2001 by Macmillan Publishers Ltd.

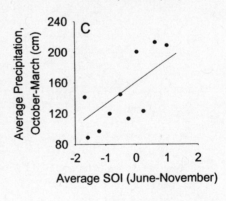

A second pattern found by Kiesecker's group was that the average depth at
which female western toads laid eggs in lakes in the Cascades was closely related to the amount of precipitation during the previous winter (Figure 4.3B).
Winter precipitation, in turn, was related to the severity of El Niño conditions the previous summer (Figure 4.3C). Climatologists have suggested that
one consequence of global climate change might be increased frequency and
intensity of El Niño conditions. Thus Figure 4.3, from the article by Kiesecker and his colleagues (2001), suggests a possible link between global climate change and declining amphibian populations: climate change may have

produced more extreme El Niño events, which caused lower snowfall in the Cascades, which caused water depth to be less around the edges of ponds where amphibians laid their eggs, which exposed the eggs to more UV radiation during development, which increased the likelihood that the eggs would be killed by a fungus. Combined with all the other sources of natural mortality on the amphibians during their life cycle, this caused populations to decline. The scenario developed by Kiesecker's group from the correlations in Figure 4.3 doesn't preclude the possibilities that thinning of the ozone layer or release of hatchery fish infected with *Saprolegnia* into high-elevation lakes in the Cascades also contributed to declining populations of western toads. But it illustrates a mechanism by which global climate change might adversely affect local populations of amphibians. Since increased UV radiation and the fungus are both necessary for large reductions in egg survival (Figure 4.2), it seems likely that population declines are a response to local factors, as well as to global changes.

My purpose in this chapter has been to show how carefully designed experiments combined with critical observations of patterns in nature can help answer ecological questions quite convincingly. Lab experiments can be useful, as in the study by Pieter Johnson and his colleagues (1999) on the effects of trematodes on limb deformities in Pacific treefrogs. Well-designed field experiments have a great deal of power to test the effects of particular factors, such as ultraviolet radiation, against a background of natural environmental conditions. And experiments in ecology, rarely persuasive on their own, require complementary information ranging from detailed understanding of the natural history of focal species to correlations between species characteristics and environmental factors. This key role of nonexperimental data was illustrated by both of the examples discussed in this chapter.

Obviously, neither limb deformities in amphibians nor widespread declines of amphibian populations are completely explained by the research of Johnson, Blaustein, Kiesecker, and their colleagues (Blaustein and Kiesecker 2002; Blaustein and Johnson 2003). For example, some declining amphibian populations in the tropics occur in densely shaded habitats; UV exposure can't be part of the explanation for these declines. Therefore, although an effect of global environmental change on amphibians in the Pacific Northwest seems likely, whether or not there is a common contributing factor to most declining populations worldwide cannot yet be answered. Final answers in science are difficult to achieve, especially in fields like ecology, which deal with very diverse and complex systems. But this contributes to the challenge of doing science, which captivates most active researchers.

RECOMMENDED RESOURCES

Blaustein, A. R., and P. T. J. Johnson. 2003. The complexity of deformed amphibians. *Frontiers in Ecology and the Environment* 1:87–94. This is a thorough review of hypotheses and evidence for limb deformities in natural populations of amphibians.

Blaustein, A. R., and J. M. Kiesecker. 2002. Complexity in conservation: Lessons from the global decline of amphibian populations. *Ecology Letters* 5:597–608. Like the article by Blaustein and Johnson on limb deformities, this review discusses the complexity of causation (see also Chapter 6), in this case, the complex causes of declining amphibian populations worldwide.

Souder, W. 2000. *A plague of frogs: The horrifying true story.* Hyperion, New York. A popular account of the discovery of frog deformities and initial research on the causes of deformities, this is especially good at introducing the personalities of the scientists involved in the research.

University of Michigan Museum of Zoology. 2003. Animal Diversity Web. http://animaldiversity.ummz.umich.edu (accessed October 1, 2003). This is a good starting point for lots of information about amphibians and many other kinds of animals.

Chapter 5

How Do Animals Find Stored Food?
Strong Inference by Testing
Alternative Hypotheses

In 1964, John Platt published a famously provocative article in the journal *Science* in which he developed a blueprint for success in science called strong inference and used molecular biology and high-energy physics to illustrate rapid progress through the systematic use of this process. According to Platt, strong inference is a process in which multiple alternative hypotheses are proposed to answer a question, and then experiments are designed to discriminate clearly among contrasting predictions of these hypotheses. If researchers are creative enough to come up with a complete set of plausible hypotheses and clever enough to design an efficient set of experiments to eliminate all but one of them, questions can be answered quickly and researchers can move on to the next step in trying to understand a phenomenon.

Platt's concept of strong inference was built on the writings of Sir Karl Popper, an influential British philosopher during the middle of the twentieth century, and T. C. Chamberlin, an American geologist who was active around 1900. Popper argued that hypotheses can never be proven true because it's always possible that a different hypothesis would lead to the same set of predictions. Therefore, scientists should focus on *disproving* hypotheses. If we are unsuccessful in disproving an hypothesis after many rigorous attempts, we may have more confidence in it, although it's impossible to ever be absolutely sure that the hypothesis is correct. The essence of Popper's (1965) argument is that attempts to disprove hypotheses are more powerful than attempts to find evidence consistent with them. Chamberlin (1890) published an article entitled "The Method of Multiple Working Hypotheses" in which he discussed the psychological benefits of considering alternative hypotheses at the same time. By not being fully committed to a single, favorite hypothesis, individual scientists would be better able to maintain objectivity about evidence relating to something they were trying to explain.

Platt's article was provocative because he contrasted the rapid progress of molecular biology and high-energy physics to the lack of progress in fields whose practitioners do not rigorously apply strong inference. Although he didn't name names, Platt was fairly blunt in his criticism of unproductive, traditional approaches: "Unfortunately, I think, there are other areas of science today that are sick by comparison, because they have forgotten the necessity for alternative hypotheses and disproof. Each man has only one branch—or none—on the logical tree, and it twists at random without ever coming to the need for a crucial decision at any point" (1964:350). Platt concluded by suggesting that government agencies that fund scientific research invest in "the man with the alternative hypotheses and the crucial experiments" rather than "the man who wants to make 'a survey' or a 'more detailed study'" (1964:352).

Ecologists and conservation biologists argued heatedly about the applicability of strong inference to their fields through the 1980s, but proponents and opponents of Platt's prescription for doing science seem to have since reached a rapprochement. Few would deny the advantages of systematic and rigorous evaluation of alternative hypotheses. At the same time, the ecological world is truly more complex and diverse than the molecular world, which means that sweeping generalizations about ecological problems, such as the causes of declining amphibian populations (Chapter 4), are less likely to exist than generalizations about the structure and function of DNA, for example. Like ecology, medical science has a kind of complexity that is qualitatively different from that of molecular biology. In ecology, this complexity comes from the tremendous diversity of species and ecosystems that exist on Earth; in medicine, the complexity comes from the fact that each individual person is genetically unique and has a unique history of interacting with his or her environment. This complexity, arising from diversity, limits the effectiveness of Platt's approach in both ecology and medicine. Nevertheless, there are examples of the successful use of strong inference in ecology and medicine, and I would like to describe one of my favorites in this chapter.

FOOD STORAGE BY CLARK'S NUTCRACKERS

Clark's nutcrackers (Figure 5.1) are members of the crow family found in conifer forests in the mountains of western North America. They depend on seeds of pines and other coniferous trees during winter, when little else is available to eat, as well as in spring and summer, when they feed these seeds to nestlings and fledglings.[1] One of the most remarkable characteristics of Clark's nutcrackers is their food-hoarding behavior. They harvest seeds and carry them up to 10 miles before burying them on wind-swept ridges and other sites that remain free of snow for much of the winter, even at high elevations. Nutcrackers have a specialized pouch under the tongue in which these seeds are carried. During the fall period of seed availability, an individual bird may

Figure 5.1. A Clark's nutcracker making a cache of pine seeds. This drawing was done by Marilyn Hoff Stewart and is reprinted with permission from *Food Hoarding in Animals* by S. B. Vander Wall, copyright ©1990 by the University of Chicago Press.

store as many as 30,000 seeds in 6,000 different caches! Caches of each individual are intermingled with those of many other nutcrackers in common storage areas several square miles in size (Vander Wall and Balda 1977).

Most of this information about the natural history of Clark's nutcrackers comes from observations by Russ Balda and his student Stephen Vander Wall at Northern Arizona University in the 1970s. Vander Wall and Balda represent a long and rich tradition in biology of those who find their motivation for doing science in a deep and abiding love of nature. As children, many have spent every available moment outdoors; when they go to college and discover that they can do the same thing in their careers, they cannot believe their good fortune. Many other factors motivate people to become scientists (see Chapter 10), but this boundless curiosity about the natural world has been one of the most productive. In fact, this story, which begins with two field biologists who are following nutcrackers around in the San Francisco Mountains of Arizona, eventually leads to remarkable new discoveries about brain structure and function in laboratories at Oxford University and elsewhere.

Vander Wall's observations of the hoarding behavior of Clark's nutcrackers led him to wonder how they relocate their caches. The supply of stored food that birds made in the fall was clearly very important both for survival over the winter and for reproduction in the next spring. The birds invested a great deal of time and energy in storing seeds in the fall, making several lengthy trips each day for several weeks from stands of piñon pine, where seeds were harvested, to the storage areas. Caches seemed to be carefully buried in the soil so they would not be readily visible. In thinking about how Clark's nutcrackers might find these buried caches, Vander Wall imagined five alternative hypotheses. This approach followed the model that Chamberlin had suggested in 1890 and that Platt had advocated in 1964 in his famous article on strong inference, although Vander Wall wasn't deliberately following the guidelines set out by these two authors. Here are Vander Wall's hypotheses:

1. *Random search:* Nutcrackers may dig randomly in storage areas and recover cached seeds by chance in the process.
2. *Directed random search:* This is a modification of the first hypothesis. Perhaps nutcrackers bury seeds in particular kinds of locations and use random search in these kinds of locations to find the caches. If so, this would be more efficient than randomly searching for caches in many different kinds of locations.
3. *Olfactory cues:* Buried seeds may emit odors or other cues that nutcrackers can detect while walking on the ground.
4. *Microtopographic cues:* Birds alter the surface of the soil when burying seeds. They find buried seeds by searching for these small-scale changes in topography produced in the process of caching.
5. *Spatial memory:* As described by Vander Wall (1982:85), this hypothesis means that "nutcrackers remember the precise location of each cache using visual cues." Since individual birds make thousands of caches, some of which they recover months later, this hypothesis implies remarkable memory ability.

Vander Wall (1982) used a series of laboratory experiments to discriminate among these hypotheses. He did the experiments with four captive Clark's nutcrackers in a large flight cage with solid walls, a wire-mesh roof, and about 3 inches of soil in which the birds could cache piñon pine seeds. He placed about 70 objects (rocks, logs, shrubs, and vertical perches) in this arena to serve as visual cues for the birds, in case they needed such cues to remember the locations of their caches (Figure 5.2). The birds moved freely around the cage in searching for food, although only two of the four subjects cached seeds during the experiments.

The key to testing the five hypotheses was to recognize that they make contrasting predictions. For example, one basic prediction of the spatial memory hypothesis is that birds should be able to recover their own caches but should not be able to find those made by other birds or by an experimenter unless they observe the caches being made. If any of the other hypotheses is correct, however, birds should be able to locate caches made by other birds or an experimenter. They might use olfactory or microtopographic cues at the cache site to do so, or they might simply use random search or directed random search.

In his first experiment, Vander Wall gave each of the caching birds (named Orange and Red, based on the colors of their leg bands) several opportunities to cache seeds over a period of about 2 weeks. After Orange had made 150 caches and Red had made 177, Vander Wall removed 50 and added 102 of his own at different locations, so there was a total of 379 available caches when he began tests of the birds' ability to locate caches. The purpose of removing some of the caches initially made by Orange and Red was to see if they would

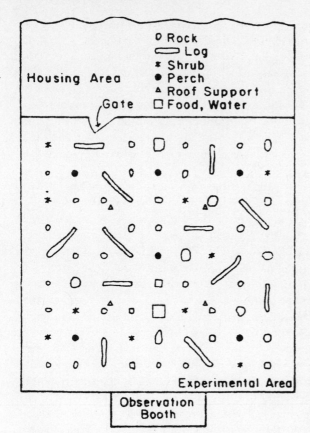

Figure 5.2. The experimental arena used by Vander Wall to study food caching by Clark's nutcrackers. Reprinted with permission from "An Experimental Analysis of Cache Recovery in Clark's Nutcracker" by S. B. Vander Wall, *Animal Behaviour* 30:84–94, copyright ©1982 by Elsevier Science.

search in those locations anyway. If they did, this would be strong evidence against the olfactory and microtopography hypotheses and in favor of the spatial memory hypothesis.

The two noncaching birds, Green and Blue, were released into the arena during some of the bouts in which Orange or Red cached seeds, but Orange never saw Red cache and vice versa. Since Green and Blue did not cache, Vander Wall wanted to see if they would be more successful at finding caches they observed being made by Orange or Red than at finding caches made by Orange or Red when Green and Blue were absent. In tests of the ability of the birds to find buried seeds, Vander Wall first let the two noncaching birds search together in the arena for a total of 6 hours over 2 days. Then the two caching birds, Orange and Red, searched individually for several hours on the next 2 days. Caches that were removed by Green or Blue on days 1 and 2 were not available to Orange and Red on days 3 and 4, so if Orange and Red were using memory to relocate their own caches, they might search in a site that was already empty.

There were dramatic differences in the searching strategies and success rates of the noncaching and caching birds. All birds probed the soil with their bills, but the noncaching birds often probed unsuccessfully near pebbles, large objects, or depressions in the surface, whereas caching birds usually went directly to cache sites before probing. Green and Blue found 52 caches in 563 probes, for an overall success rate of about 9%, whereas Orange and Red found 52 caches in 140 probes, for a success rate of 37%. However, many of the unsuccessful probes of Orange and Red were at sites where caches had already been removed, either by the experimenter before the recovery tests or by Green or Blue. If these probes are counted as successes, since the birds were searching where there should have been caches, the overall success rate increases to 71%.

Orange found 63 of its own cache sites but none of those made by Red, whereas Red found 61 of its own cache sites and three made by Orange. Neither of these caching birds found any of the caches made by Vander Wall himself. By contrast, the two noncaching birds found 11 caches that had been made by Orange, 30 that had been made by Red, and 11 that had been made by Vander Wall. Both the foraging behavior of the birds and their patterns of success imply that the caching birds were using different strategies for finding buried seeds than the noncaching birds. The results strongly suggest that spatial memory was used by the caching birds to relocate their own caches. For the noncaching birds, spatial memory was evidently not important because these birds found more of the caches made by Orange and Red when their caching was *not* observed by Green and Blue than when caching was observed. Green and Blue also found several caches made by the experimenter, who had placed these caches without being seen by the birds.

Vander Wall used the results for the noncaching birds to discredit the random search hypothesis, as well as the spatial memory hypothesis. This required calculating the probability of detecting a cache by probing in a random location and comparing this to the success rate of the noncaching birds. At the beginning of the recovery portion of the experiment, there were 379 caches in an area of about 75 square meters. The area of each cache was about 2 square centimeters, and the area excavated in a probe was also about 2 square centimeters. Therefore the probability of hitting any part of a cache with one probe was about 0.4%.[2] This probability is much lower than the actual 9% success rate of the noncaching birds, implying that these birds were either using some specific cues at cache sites or limiting their search to areas in the arena that had higher densities of caches.

Vander Wall designed a second experiment to test the hypothesis that Clark's nutcrackers use microtopographic cues to help locate caches. After completion of the first experiment, the two caching birds made additional caches in the arena, and Vander Wall made some caches himself. He then carefully raked one half of the arena to erase any marks on the soil surface that might have been made either by the birds or by himself while caching seeds. If microtopographic cues are important in locating caches, the birds should have been more successful in the unraked control portion of the arena than in the raked experimental portion.

As in the first experiment, the two noncaching nutcrackers were allowed

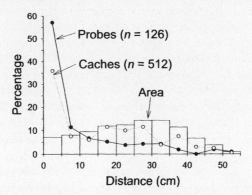

Figure 5.3. Locations of caches and probes in search of caches made by Clark's nut-crackers in relation to large objects (see Figure 5.2) in Vander Wall's arena. The open circles show the percentages of caches at various distances from large objects: 0 to 5 centimeters (cm), 5 to 10 cm, and so on. The solid circles show the percentages of probes for caches at these distances. The bars show the percentages of the total area of the arena at these various distances from large objects. Modified with permission from Figure 3 in "An Experimental Analysis of Cache Recovery in Clark's Nutcracker" by S. B. Vander Wall, *Animal Behaviour* 30:84–94, copyright ©1982 by Elsevier Science.

to search for caches before the two caching birds. One of them had a higher proportion of successful probes on the control portion of the arena than on the experimental portion, but the other did not. However, both were more successful at finding recently made caches (less than 3 days old) than older caches (6 to 24 days old). They found 13% of recent caches and 3.4% of older caches in the control area. This difference suggests that the birds used micro-topographic cues at cache sites to help locate caches because these cues would degrade with time, making older caches less apparent than more recent ones.

The results of this experiment provided some support for the directed random search hypothesis, as well as the microtopographic cues hypothesis. Figure 5.2 shows the array of objects, such as logs and rocks, that Vander Wall placed in the arena to make it more similar to a natural environment. The two caching nutcrackers made most of their caches within 20 centime-ters of these objects, and the two noncaching birds did most of their digging close to the objects as well (Figure 5.3). Vander Wall used the results in Fig-ure 5.3 to estimate a predicted success rate of 1.8% if noncaching birds were using directed random search in experiment 2, which was similar to the over-all success rate of 1.6% for these two birds in the whole arena.

After the two noncaching birds searched for buried seeds in the arena in experiment 2, the birds that had made the caches were given an opportunity to do so also. Their proportion of successful probes was quite a bit lower in the experimental half of the arena than in the control half, suggesting that *raking the soil surface had affected their searching behavior.* Just as in experi-ment 1, they probed at several sites where caches had been removed by the noncaching birds. Vander Wall's detailed observations hinted that the birds might have been confused by the discrepancy between their memory of specific cache locations and the absence of microtopographic cues at these locations in the experimental half of the arena: "Birds would frequently approach a cache site and look at the disturbed [raked] soil surface while hopping from side to side, backwards or in a small circle. The head was occasionally turned toward large objects. After one or two probes near the cache site, the seeds were usually found" (1982:89).

The results of the first two experiments were inconsistent with the hypothesis of random search because success rates were much higher than they would have been if based solely on random search. Noncaching birds appeared to rely on searching near objects where caches were usually made (directed random search) and on microtopographic cues to locate recent caches made by caching birds. Caching birds used spatial memory to relocate their own caches and attended to microtopographic cues, although these weren't essential. There was no evidence that olfactory cues emanating from the caches themselves were used by the birds because the caching birds did not dig up caches made by others and because raking half of the arena decreased the ability of noncaching birds to locate caches (raking eliminated microtopographic cues but shouldn't have affected olfactory cues). With this background, Vander Wall designed a third experiment to further test the spatial memory hypothesis. This was the kind of clever, critical experiment that Sir Karl Popper and John Platt would have appreciated because it provided a conclusive test of the hypothesis.

Vander Wall used just the two caching birds in the third experiment. After allowing each of them to make caches in the arena, he moved all large objects in one half of the arena 20 centimeters (about 8 inches) to the right. He then smoothed out the depressions where these objects had been but avoided disturbing the caches or the surface of the soil. As a control, objects in the other half of the arena were not moved. The birds were much more successful at finding caches in the control portion of the arena (73% of probes were at cache sites) than in the experimental portion (16% of probes were at cache sites). More important, 69% of the errors in the experimental portion of the arena were about 20 centimeters to the right of cache sites, whereas only about 11% of errors in the control portion of the arena were similarly displaced. Since Vander Wall had moved the large objects in the experimental half of the arena 20 centimeters to the right after caching but before recovery, this pattern of errors strongly supports the hypothesis that the nutcrackers were using these objects as visual cues to help them remember the locations of their caches.

STRONG INFERENCE AND INDIVIDUAL VARIATION IN BEHAVIOR

You may be skeptical of this evidence that Clark's nutcrackers are able to use spatial memory to relocate seed caches since Vander Wall tested only four birds and only two of these made caches in the arena. Researchers are often concerned about having adequate sample sizes to test hypotheses rigorously, but the importance of this depends on the nature of the hypothesis and the outcome of the experimental test. For example, Chapter 8 discusses tests of the hypothesis that caffeine causes elevated blood pressure in humans. There was a great deal of individual variation in the responses of subjects to caffeinated versus decaffeinated coffee, so researchers found it necessary to test a fairly large number of subjects in order to get precise estimates of effects on blood pressure. The problem was different in this example of food hoarding by Clark's nutcrackers because Vander Wall was testing a different kind of hypothesis: that birds could find caches by using spatial memory. In the blood pressure example, there was a continuous range of possible responses to caffeine, and the problem was to test enough subjects to get an accurate and precise estimate of the average response. In the cache recovery example, the response was dichotomous: either birds could use spatial memory to find caches or not. If Vander Wall could clearly show that even one bird had this specialized and sophisticated cognitive ability, he would have produced interesting new knowledge.

There were two potential pitfalls in Vander Wall's study with only four birds. If none of them had cached in captivity, he couldn't have tested the hypothesis. If some of the birds had cached but there was no evidence that they used spatial memory to relocate caches, this would not be strong grounds for refuting the spatial memory hypothesis because the artificial conditions of the experiment might have impeded this ability. In fact, two birds did cache and several aspects of their cache recovery behavior were consistent with the spatial memory hypothesis. The most impressive match between the results and the hypothesis was that the errors made by the birds when large objects in the arena were shifted were precisely what would be expected if the birds had been using these objects as landmarks.

It's worth comparing these experiments to the study described in Chapter 3, in which Schoon (1998) tested the ability of trained dogs to identify odors in police lineups that matched odors left at crime scenes. In Schoon's experiment, some dogs didn't perform adequately in preliminary trials, suggesting that they weren't motivated to perform in the trials that counted. These were comparable to the two birds in Vander Wall's experiment that didn't cache in captivity. But even the dogs that passed the preliminary trials in Schoon's study were not very successful in the trials that counted. More important, Vander Wall designed his experiment so that the pattern of errors, as well as the overall success rate under different conditions, could be used to test the spatial memory hypothesis. Therefore, Vander Wall was able to reach a definitive conclusion about the spatial memory abilities of his subjects, whereas in Schoon's case the best we could say is that different training methods or a different experimental design might validate the use of dogs in forensic identification but that current evidence doesn't support this use.

It seems clear that at least two Clark's nutcrackers can use spatial memory to find their caches. Perhaps these were very unusual birds, and Vander Wall was incredibly lucky that two of the four birds he captured for his experiment had this ability. This seems highly unlikely, especially considering the fact that Clark's nutcrackers make thousands of caches in nature, that these caches are interspersed with those made by other individuals, and that they are recovered months later. This is a situation in which spatial memory would be very valuable, so it's not plausible that the ability would be limited to just the two birds that Vander Wall happened to capture for his experiments.

This example differs in three important respects from Platt's (1964) description of strong inference based on the history of molecular biology and high-energy physics that I described at the beginning of this chapter. First, the hypotheses considered by Vander Wall to explain cache recovery by Clark's nutcrackers were not mutually exclusive. Second, Vander Wall found significant individual variation among four birds in how caches were found. Two birds, which did not make their own caches, relied on directed random search and microtopographic cues to find caches made by others. Two birds, which did cache, relied primarily on spatial memory but were also influenced by microtopographic cues.[3]

These first two differences between Platt's analysis and Vander Wall's work reflect the complexity of problems in ecology (and medicine) compared to those in molecular biology and physics. Despite this complexity, Vander Wall successfully used critical experiments to compare alternative hypotheses. This strong inference approach not only advanced the understanding of food hoarding but also produced the first definitive evidence that some birds have impressive spatial memory abilities.

The third difference between Vander Wall's work and Platt's description of strong inference was more fundamental. Platt's version of strong inference was based on Popper's (1965) philosophy of falsificationism, in which hypotheses can never be "proven" true but only falsified. In Platt's view, a productive approach in science is to consider a set of alternative, mutually exclusive hypotheses; design experiments to falsify each of them; reject all but one of the hypotheses based on the results of these experiments; and build a new set of hypotheses to refine the one that could not be rejected. Vander Wall's experiments enabled him to reject some of his hypotheses about how Clark's nutcrackers locate cached seeds, but also provided strong *positive* evidence for the spatial memory hypothesis. This approach—not simply trying to falsify hypotheses but also doing experiments to confirm hypotheses—seems to be more common in science than Platt or Popper would like, and Vander Wall's work shows that it can be very productive.

OTHER TESTS OF SPATIAL MEMORY IN FOOD-HOARDING ANIMALS

Vander Wall's pioneering studies of Clark's nutcrackers have been extended in several ways. Russ Balda, Alan Kamil, and their students have continued to study Clark's nutcrackers and related species of jays in large enclosures. In one experiment, they asked how long Clark's nutcrackers could remember locations of cached seeds (Balda and Kamil 1992). They used 25 birds, each of which made approximately 21 caches in an indoor arena with a surface area of

30 by 50 feet. The birds cached in sand-filled cups placed in 330 holes drilled in the plywood floor of the arena. Each bird cached individually and was tested with only its own caches in place. As in Vander Wall's experiment, Balda and Kamil placed numerous large objects in the arena to serve as landmarks. Following caching, the birds were divided into four groups to be tested for spatial memory at 11 days, 82 days, 183 days, or 285 days after the caches were made.

Balda and Kamil found that the birds in their experiment were more successful in recovering caches than would be predicted by random search at all time intervals, including 285 days, although birds that had to remember cache locations for 285 days made more errors than birds in the other three groups. The long-term spatial memory demonstrated by Balda and Kamil in this experiment is consistent with the natural history of Clark's nutcrackers, which harvest some caches in spring and summer that were made in the previous autumn. This experiment also illustrates some refinements in technique over Vander Wall's earlier experiment. Balda and Kamil tested many more birds, all of which cached. They used a larger arena and provided small cups as specific sites for caching rather than allowing them to cache anywhere in the entire arena. The latter setup, which had been used by Vander Wall, was criticized on the grounds that birds might not remember specific cache locations but instead would consistently cache at specific distances and directions from objects in the environment and then search at those sites when they were recovering caches. The design of Balda and Kamil precluded this possibility because the cups in which birds were able to cache were at variable distances and directions from landmarks in the arena. This design has been used in most subsequent laboratory studies of food hoarding by other species.

Various researchers have shown in more than 100 studies that several other species of birds and mammals use spatial memory to recover their food caches (Krebs 1990; Shettleworth 1990). These species include other members of the crow family such as pinyon jays, scrub jays, gray jays, and northwestern crows; members of the tit family such as black-capped chickadees in North America and marsh tits in Europe; and seed-eating rodents such as gray squirrels, chipmunks, and kangaroo rats.[4] Brief description of a few studies will give you the flavor of this research. Balda and Kamil (1989) compared the spatial memory ability of three species of jays and Clark's nutcrackers. They predicted that pinyon jays and Clark's nutcrackers would have better spatial memory than scrub jays and Mexican jays because the former two species spend winters at higher elevations in harsher climatic conditions and are more dependent on cached pine seeds to survive. Under standard laboratory conditions, the researchers found that pinyon jays and Clark's nutcrackers were indeed more successful at remembering locations of their caches than scrub jays and Mexican jays.

Researchers have used observations and experiments in the field, as well as laboratory experiments in arenas, to test spatial memory abilities of animals. For example, Hutchins and Lanner (1982) reported that Clark's nutcrackers dig directly through snow to recover caches, suggesting that they use remembered landmarks rather than local microtopographic cues, which would be obliterated by snow cover. Stevens and Krebs (1986) did a clever field experiment to test the spatial memory of marsh tits near Oxford in England. They placed leg bands with tiny magnets on birds and then allowed the birds to harvest seeds from a feeder and cache them. Then they positioned magnetic detectors connected to small clocks near each of 135 caches and 60 control sites where the birds had not stored seeds. In 3 days, the detectors recorded 32 visits to cache sites and no visits to control sites. During the first 12 daylight hours after storage, the marsh tits recovered 91% of the seeds at caches they visited. After that time, their success rate declined to 20% because many caches had been stolen by rodents or other birds.

Just as two of the nutcrackers in Vander Wall's experiment used microtopographic cues to find caches made by other birds, rodents use olfaction, as well as spatial memory, to locate caches. Mammals in general and rodents in particular have more acute senses of smell than most birds, at least for seeds, and several researchers have shown that some rodents can use olfaction to find seeds buried several centimeters deep in the soil. This means that caches made by birds and rodents are vulnerable to pilferage by other rodents. However, Vander Wall (1998) found in a series of laboratory and field studies that the ability of rodents to locate buried seeds by smell depends on the moisture content of the soil. Immediately following rain, caches may be very apparent to several species of rodents; as the soil dries out, caches become difficult to smell. Rodents such as kangaroo rats that are adapted to arid environments require less soil moisture to smell buried seeds than rodents such as chipmunks that live in less arid environments. Vander Wall (2000) has used a double-labeling technique, in which pine seeds are labeled with a radioactive isotope, so they can be found in the soil by a Geiger counter, and with an indelible number, so they can be identified individually, to show that some rodents continuously rearrange their caches. For example, one chipmunk might use smell to find a cache made by another animal, then dig it up and rebury it elsewhere, giving it an extra edge in recovering it later through spatial memory.

SPATIAL MEMORY AND THE BRAIN

In discussing how animals relocate their food caches, I have focused on Vander Wall's experiments with Clark's nutcrackers because they illustrate the use of strong inference and creative thinking in designing a critical experiment to discriminate among alternative hypotheses. You've seen several examples of the experimental approach in Chapters 2–4, but this food-hoarding example is especially attractive because it provides persuasive evidence of the use of spatial memory by Clark's nutcrackers. Vander Wall's experiments were not perfect, but they were convincing enough to persuade many other researchers to do follow-up studies of Clark's nutcrackers and other species that showed beyond reasonable doubt that some animals have impressive abilities in the use of spatial memory.

Some kinds of questions aren't as amenable to experimental approaches as questions about foraging behavior and spatial memory abilities. For example, species such as Clark's nutcrackers and other kinds of jays differ in how much they cache seeds and in spatial memory ability. Are there corresponding differences in the brains of these species? If so, these neurological differences could be interpreted as adaptations to different environmental challenges that have evolved over many generations. But the evolution of these adaptations cannot usually be directly studied through experimentation because it happens over very long periods of time. Instead, evolutionary biologists use an approach called the *comparative method.*

Comparisons are a fundamental tool in testing hypotheses in both experimental and purely observational research. Experiments, for example, almost always involve comparison of a treatment and control group, as illustrated in all of the chapters so far. When the comparative method is used in nonexperimental research, comparisons are made between different individuals of the same population, different populations of the same species, or different species—with no experimental manipulations applied to these entities. Hypotheses about adaptation are commonly tested by seeing if differences in one trait are related to differences in another trait or to differences in the environments where the individuals, populations, or species live. For example, Eisenberg and Wilson (1978) measured the brain sizes of 225 species of bats. Not surprisingly, they found that species with larger average body size tended to have larger brains. However, when they divided their sample into insect-eating bats and fruit-eating bats, they found a more interesting pattern: fruit-eating bats had larger brains than insect-eating bats of similar body size. They suggested two possible explanations: (1) fruit-eating bats might benefit from remembering where concentrations of ripe fruit occurred in their home ranges at different times of year, requiring enlarged brains, or (2) insect-eating bats might benefit from *minimizing* the sizes of their brains so they could maneuver more easily while flying and thus have greater success at catching insects. This correlation of a morphological characteristic of various species of bats (brain size) with a behavioral trait (food habits) led to two evolutionary hypotheses about the possible advantages (and disadvantages) of large brains that could be tested by more detailed comparative studies.

Platt's (1964) program of strong inference relies heavily on devising and conducting critical experiments to discriminate among alternative hypotheses. Does this mean that strong inference is limited to fields in which experimental tests of hypotheses are feasible? I think not. One way to marry strong inference and the comparative method is to use multiple comparisons of different groups of animals in different situations to test a common set of hypotheses. Although any one of these comparisons might be compromised by the possibility of an alternative explanation, it's unlikely that all of the comparisons would be compromised in this way. Therefore, if different comparative studies are consistent with the same hypothesis, this provides strong evidence for the hypothesis, especially if the comparative studies are quite

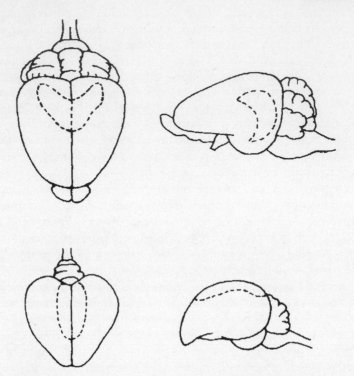

Figure 5.4. Dashed lines show the position of the hippocampus in the brain of a red squirrel (top) and a chickadee (bottom). The drawings on the left show top views, with the front of the brain at the bottom; those on the right show side views, with the front of the brain at the left. The brains of the two species are not drawn to the same scale. Reprinted with permission from "Neurobiological Bases of Spatial Learning in the Natural Environment: Neurogenesis and Growth in the Avian and Mammalian Hippocampus" by D. W. Lee, L. E. Miyasato, and N. S. Clayton, *NeuroReport* 9:R15–R27, copyright ©1998 by Lippincott Williams & Wilkins.

diverse. I'll illustrate this by describing several tests of a recent hypothesis about neurological foundations of spatial memory.

Birds and mammals have a distinctive region of the brain, called the hippocampus, that is involved in storing and using spatial memories (Figure 5.4). It is shaped roughly like a sea horse, which the Greeks called *hippokampos*. Until about 1980, brain researchers universally believed that all nerve cells in the brain form during prenatal or early postnatal development and that no new nerve cells form in adult brains. In 1981, however, Fernando Nottebohm showed that new nerve cells are produced each spring in the region of canary brains that controls song learning during the breeding season. Although this process of nerve cell regeneration doesn't occur in most of the brain, the hippocampus is another region in which new nerve cells can be produced throughout life. This suggests that individuals who make greater use of spatial memory might develop larger hippocampi.

A study of cab drivers by Eleanor Maguire and her colleagues (2000) at the University College of London may illustrate this relationship. Maguire's group studied cabbies in London who train for approximately 2 years before taking several challenging exams, which test their knowledge of thousands of locations and optimal routes between them at various times of the day. The researchers used a procedure called magnetic resonance imaging (MRI) to measure the size of the hippocampus in 16 male taxi drivers and 50 other men. A specific portion of the hippocampus was larger on average in the taxi drivers than in the men with other professions. In addition, taxi drivers with more experience had larger hippocampi than drivers with less experience, consistent with the hypothesis that extended use of spatial memory promotes cell division in the hippocampus. Another technique, the positron emission tomography (PET) scan, allows visualization of the parts of the brain that are active during a particular task. This showed a concentration of activity in the rear of the hippocampus when taxi drivers were figuring out the best routes between sites in London.

This study of cab drivers illustrates a comparative approach to differences among individuals belonging to the same population. The researchers tested the hypothesis that humans can respond to unusual demands on their spatial memory by adjustments in the size and functioning of a part of the brain thought to play a key role in storing and using spatial memories. The results are intriguing, although far from conclusive, because there are many other differences between cab drivers and other people in the same city besides the fact that the former have to rely more on spatial memory than do most people. Cab drivers might also be more stressed or exposed to higher levels of carbon monoxide or smog than the average person, which could affect their brains (Nicola Clayton, personal communication). This is another illustration of the difficulties of interpreting correlational data, which I introduced in Chapter 2.

If food-storing animals have well-developed spatial memory abilities, they should also have larger hippocampi than closely related species that don't store food. Several researchers have tested this prediction, including John Krebs at Oxford University, David Sherry at the University of Toronto, and Lucia Jacobs at the University of California at Berkeley. The basic technique in these studies was to dissect the brain; cut it into very thin cross sections; stain these sections with a dye that highlights nerve cells; identify the boundary of the hippocampus on each section, based on the size and type of cells present; measure the area of the hippocampus in each section; and convert area to volume by accounting for the width of each section. Krebs and Sherry (Krebs et al. 1989; Sherry et al. 1989) studied the brains of many different species of birds. Since these species differed greatly in body size and in total size of the brain, the researchers compared the species in terms of the *relative* size of the hippocampus, defined as the proportion of the forebrain that was occupied by the hippocampus (the forebrain, one of the major sections of the brain, handles sensory input and controls muscular activity). In making this comparison, the researchers were essentially asking how much brain tissue was allocated to spatial memory tasks, performed in the hippocampus, compared to other tasks performed in other parts of the forebrain.

Krebs, Sherry, and their colleagues found that the relative size of the hippocampus was larger in food-storing birds than in species that do not store food. The most informative comparisons were for closely related species with similar behavior and ecology except for the use of food hoarding. For example, marsh tits weigh about 11 grams and have a hippocampus that occupies about 14.3 cubic millimeters, whereas great tits weigh 20 grams and have a hippocampus of only 11.2 cubic millimeters. The whole brain of great tits is quite a bit larger than that of marsh tits, in keeping with the larger body size of the former. In marsh tits, which store food, the hippocampus is 5.3% of the volume of the forebrain. In great tits, which do not store food, the corresponding value is 3.3%. In other words, the relative size of the hippocampus is about 60% greater in marsh tits than in great tits. Similar differences exist for storing and nonstoring members of the crow family.

In all of the food-storing bird species that have been studied neurologically, individuals scatter hoard by making numerous caches at widely dispersed locations in their home ranges. In small mammals, by contrast, there are two major patterns of food hoarding: scatter hoarding and larder hoarding, in which items are stored in a central location in the home range, often a large burrow. Jacobs and Spencer (1994) took advantage of this diversity to make a comparison with an interesting twist. Merriam's kangaroo rats are primarily scatter hoarders, and bannertailed kangaroo rats are larder hoarders. This may be related to the fact that the latter are about three times as large as the former and can defend seeds stored in their burrows against pilferage by the smaller species. The relative volume of the hippocampus is about 20% larger in Merriam's kangaroo rats than in bannertailed kangaroo rats, consistent with the fact that the former would be more dependent on spatial memory to relocate their numerous, scattered caches.

Marsh tits and great tits have had separate evolutionary histories for a very long time, as have Merriam's kangaroo rats and bannertailed kangaroo rats, so these results are consistent with the hypothesis that larger hippocampi have evolved in species that make greater use of spatial memory in their foraging behavior. However, species with relatively large hippocampi may differ from related species with smaller hippocampi in several behavioral traits besides scatter hoarding, so the interpretation that a large hippocampus is adapted to the spatial memory requirements associated with recovering stored food is not foolproof. Support for this hypothesis is strengthened, however, by the fact that a relationship between hippocampus size and food-hoarding behavior was found in various types of birds, as well as in kangaroo rats. It seems unlikely that a parallel set of behavioral differences, independent of food hoarding, could exist between species with large and small hippocampi, given the wide range of variation in behavior that exists between the different bird species studied (jays and tits) and between these birds and kangaroo rats. For example, all of the bird species are active in the daytime, whereas the kangaroo rats are nocturnal; many of the bird species are monogamous, whereas the kangaroo rats are not monogamous; and so on. In addition, the comparison for kangaroo rats was between two species that store food in different ways, not between storing species and nonstoring species, as in the bird studies. The kangaroo rat results imply that there is a specific association between the size of the hippocampus and a type of food storage that would benefit from good spatial memory, not simply a generalized relationship between the size of the hippocampus and the presence or absence of food hoarding.

The association between an enlarged hippocampus and enhanced spatial memory was reinforced by a recent study by Vladimir Pravosudov and Nicola Clayton (2002) at the University of California at Davis in which they compared two populations of black-capped chickadees, one from Alaska and one from Colorado. They argued that food caching would be more important for winter survival in the harsher environment of Alaska than in Colorado, where food is presumably more abundant and more predictably available. Therefore, they predicted that chickadees from Alaska would cache more food, would recover caches more efficiently by using spatial memory, and would have larger hippocampi than chickadees from Colorado. They tested these predictions with captive birds from the two environments kept in the same conditions in the laboratory, and their results were consistent with all three predictions.

The hippocampus has been the focus of other fascinating comparative research in recent years. This research broadens support for the hypothesis that situations in which good spatial memory would be advantageous promote the evolution of larger hippocampi. Steven Gaulin and his colleagues studied two small rodents called meadow voles and pine voles (Gaulin and Fitzgerald 1986; Jacobs et al. 1990). Meadow voles have a polygynous mating system, in which one male mates with several females (and some males don't mate at all). Males have larger home ranges than females, and one male's home range usually overlaps that of several females. This is a typical pattern of mating and space use in rodents. By contrast, pine voles have a monogamous mating system, and each male-female pair shares a small home range. These differences between the two species suggested to Gaulin and his coworkers that male and female pine voles might have similar requirements for spatial memory ability, whereas male meadow voles might rely on spatial memory to a greater extent than females in order to navigate their larger home ranges and keep track of when the different females living there were receptive to mating. Indeed, male meadow voles have larger hippocampi than female meadow voles, but male pine voles have the same size hippocampi as female pine voles. Furthermore, male meadow voles learn to navigate a standard laboratory maze more quickly than female meadow voles, but there is no difference in learning ability between male and female pine voles.

Just so you don't make a premature generalization about sex differences in the size of the hippocampus based on the meadow vole results, the pattern is reversed in shiny cowbirds. These birds are nest parasites, which lay eggs in the nests of several other species of birds. Suitable host nests are often well hidden, so relocating them after an initial survey of the possibilities may require good spatial memory. In shiny cowbirds, only females select potential host nests, and females have larger hippocampi than males. In screaming cowbirds, however, both females and males inspect potential host nests, and there is no difference between sexes in the size of the hippocampus (Reboreda et al. 1996). Thus whether there are sex differences in the size of the hippocampus and, if so, whether males or females have larger hippocampi appear to depend on how species use spatial memory. All of these comparative studies of different species, different populations of the same species, and different sexes within a species are consistent with the hypothesis that hippocampi are larger in cases in which requirements for good spatial memory are greater. This diverse evidence illustrates how comparative research can produce strong inference even when experimental approaches aren't feasible.

CONCLUSIONS

In this chapter I considered various aspects of spatial memory in animals, ranging from behaviors linked to the use of spatial memory in food-storing animals to a region of the brain associated with that ability. I discussed in detail a study of Clark's nutcrackers by Stephen Vander Wall that illustrated the advantages of considering multiple alternative hypotheses and designing critical experiments to discriminate among them. I also discussed more briefly some applications of the comparative method to help elucidate evolutionary relationships between brains and behavior. Although it is often more difficult to design and execute critical experiments to test alternative hypotheses in ecology and medicine than in such fields as molecular biology and high-energy physics, there are many other good examples besides the study by Vander Wall. There are also numerous examples of applications of the comparative method to studying adaptation, many of which provide inferences about alternative hypotheses that are just as strong as in well-designed experiments.

Besides illustrating these two fundamental methods of testing hypotheses by experimentation and by systematic comparison of different species, I used this chapter to tell a story about how scientific understanding develops. The story began with basic observations of the natural history of Clark's nutcrackers, which led to a question about how they relocate the thousands of caches they make each fall. This question led Vander Wall to design a clever but relatively simple set of experiments to test five potential answers to the question. The story continued as researchers realized that food-hoarding animals would be good models for studying various aspects of spatial memory, including the neurological basis for memory. This led to comparative studies of the size of the hippocampus in species that differ in their food-storing behavior and eventually to comparisons linked to other kinds of behavioral variation. Science often follows such a tortuous path, with each new discovery leading to new questions that may take researchers in exciting new directions.

RECOMMENDED READINGS

Chitty, D. 1996. *Do lemmings commit suicide? Beautiful hypotheses and ugly facts.* Oxford University Press, New York. This is an enjoyable and stimulating history of the frustrations of trying to use strong inference to solve a complex ecological problem, determining the causes of the remarkable population cycles of lemmings.

Christie, M. 2000. *The ozone layer: A philosophy of science perspective.* Cambridge University Press, Cambridge. Christie analyzes the history of research to discover the reason for the hole in the ozone layer that develops over Antarctica each year. She shows the importance of a critical experiment in providing positive evidence to solve this problem.

Jacobs, L. F. 1996. Sexual selection and the brain. *Trends in Ecology & Evolution* 11:82–86. This is a good introduction to some recent work on the hippocampus.

Platt, J. R. 1964. Strong inference. *Science* 146:347–353. A classic provocative article about doing science effectively.

Vander Wall, S. B. 1990. *Food hoarding in animals.* University of Chicago Press, Chicago. The bible of research on food hoarding.

Chapter 6

What Causes Cancer?
The Complexities of Causation

On 13 July 2000, the *Washington Post* described a new study of cancer incidence in twins in a front-page story with the headline "Cancer Study Downplays Role of Genes." On the same day, the *New York Times*'s headline was "Genes May Cause 25% of 3 Major Cancers." Surprisingly, these two stories were based on the same technical report by a group of Scandinavian scientists that appeared in the *New England Journal of Medicine* on that day. What can we make of these different interpretations of the same scientific study? A careful reader of the *Washington Post* might conclude that environmental factors have broader scope in causing cancer than suggested by well-established examples such as smoking and lung cancer. A conscientious reader of the *New York Times* might conclude just the opposite. And a masochistic reader of both newspapers might be left with a good deal of confusion. A detailed look at this example will illustrate several aspects of the scientific process. This chapter won't completely alleviate the possible confusion by providing a definitive answer to the question of what causes cancer. However, I hope to use the multiple meanings of this seemingly simple question to illuminate a fundamental challenge of doing science, as well as one source of the enchantment of science.

The twin study by Paul Lichtenstein of the Karolinska Institute in Stockholm and his Swedish, Danish, and Finnish collaborators (Lichtenstein et al. 2000) was newsworthy because it included data on 44,788 pairs of twins, four times as many as any previous analysis of cancer incidence in twins. The researchers were able to amass such a large amount of data because of the detailed genealogical and medical records that have been maintained in these Scandinavian countries over a long period of time. For example, part of their sample included 10,503 pairs of twins of the same sex, recorded in the Swedish Twin Registry, who were born in Sweden between 1886 and 1925 and were still alive in 1961.

Data on twins are appealing for researchers interested in distinguishing the roles of environmental and hereditary factors in disease because of the possibility of comparing incidences of various diseases in monozygotic (identical) and dizygotic (fraternal) twins.[1] Because monozygotic twins share 100% of their genes and dizygotic twins share 50% of their genes, a disease with a strong hereditary component should be more common in monozygotic twins of afflicted individuals than in dizygotic twins of afflicted individuals. In fact, debate about the relative importance of nature versus nurture in causing cancer has a long and rancorous history, though not as rancorous as the parallel debate about nature, nurture, and human intelligence. In the case of cancer, the new study by Lichtenstein and his colleagues had the potential to make a significant contribution toward resolving this debate. However, their analyses and interpretations were strongly challenged by Neil Risch (2001), Colin Begg (2001), and others. These critiques were not reported in the popular media, but the nature-nurture controversy continues, at least behind the scenes.

SOME COMPLEXITIES OF CAUSATION

Before we can try to make sense of these different interpretations of incidences of cancer in twins and the implications of this work for the relative roles of genetic and environmental factors, we need to consider some historical and philosophical background. One way to answer the question "What causes cancer?" is to say that basic research in cellular and molecular biology over the past 25 years has established beyond the shadow of doubt that the cause of cancer *is* genetic. This bold statement is based on work in hundreds of laboratories around the world that has revealed fundamental differences between cancer cells and normal cells and produced an understanding of the mechanisms that cause normal cells to become cancerous. To make a long story very brief, the cause of cancer at the molecular level is mutations in DNA that disrupt the mechanisms that normally regulate cell division and growth. Several of these mutations are typically required to convert a lineage of normal cells into a full-fledged tumor capable of growing aggressively and spreading to other tissues. Most cancers are associated with old age because the development of cancer is a multistep process, with each step dependent on a mutation that occurs with low probability. Remarkably, most malignant cancers probably originate from a single cell that sustains an initial mutation, setting the lineage that arises from that cell on a path of increased cell division and eventual uncontrolled growth (Varmus and Weinberg 1993; Weinberg 1996).

Much of the mystery surrounding cancer has been eliminated through the development and clever application of new techniques in cellular and molecular biology. However, the molecular answer to the question of what causes cancer is not completely satisfying. For example, this mechanistic model of cancer doesn't explain why some individuals get cancer and others do not. Smoking greatly increases the risk of developing several kinds of cancer, but not all smokers get cancer. Is smoking more likely to trigger cancer in some individuals because they inherit a susceptibility to carcinogens in tobacco? Are there other factors in the environments of smokers that act in combination with tobacco smoke to cause cancer? Or are smokers who don't develop cancer just lucky, like lottery winners? Answering these kinds of questions opens up more opportunities for prevention and treatment of cancer than simply understanding the molecular basis for the disease.

It appears that the causes of cancer are more complex than suggested by our simple-sounding question. However, the reasons for this are more fundamental than the fact that cancer is a complex, multifaceted set of at least 120 diseases, involving different types of cells and having different treatments and likely outcomes. The same kinds of complexities arise in considering the causes of most biological phenomena. In fact, the nature of causation has been a major concern of philosophers for centuries. I won't outline this long history but rather will make some general points about how scientists think about causation before I return to the twin study. These general points will also be useful for other examples in later chapters.

A common working assumption in science is that every effect has a single cause. One reason that practicing scientists frequently make this assumption is that it's easier to think about discrete, single causes of phenomena than about complex, multiple, interacting causes. This assumption has also been of great practical value because it has stimulated the design of critical studies to test specific, focused hypotheses about causation. One good general example is the use of Koch's postulates to test the hypothesis that a particular disease is caused by an infectious organism. Robert Koch was a leading microbiologist of the nineteenth century who proposed the following criteria for establishing infectious causation of disease: (1) the hypothetical infectious agent should be isolated from individuals with the disease, (2) the infectious agent should be grown in pure culture, (3) the culture should be shown to cause disease in humans or laboratory animals, and (4) the infectious agent should be isolated from these subjects. Early researchers used this protocol with great success to discover the microbial agents that cause diseases like anthrax, tuberculosis, and syphilis (Carter 1985). However, this approach depends on the assumption that a particular microorganism is the single cause of the disease.

The assumption that a phenomenon has a single cause has two components: that the purported cause is *necessary* for the phenomenon to occur and that the purported cause is *sufficient* for the phenomenon to occur. Thus there are two fundamental ways in which causation may be more complex than is often assumed. First, a particular factor may be necessary but not sufficient to produce an effect. This means that if the factor is not present the effect will not occur, but other factors are also required to produce the effect. For example, cancer typically requires multiple mutations of different regions of DNA in the cell, including mutations of genes that promote cell division and growth and of genes that normally suppress division and growth. The second possibility for multiple causation is that a factor may be sufficient but not necessary to produce an effect. This means that the presence of the factor guarantees the effect (the factor is sufficient) but its absence does not (other factors can also produce the effect). For example, acute intermittent porphyria may be caused by an inherited genetic condition that interferes with the production of hemoglobin, the protein that carries oxygen in the blood. Its most common symptoms are anemia, intermittent but extreme abdominal pain, and mental disturbance. More commonly, however, the disease is caused by lead poisoning, which affects a different step in the process of hemoglobin synthesis.

Another way in which multiple factors may influence phenomena is that individual factors may have partial effects, so the total response is stronger if two or more factors are present than if only one is present. Risk factors that influence diseases often follow this model. For example, smoking, diet, stress, high blood pressure, and lack of exercise all increase the risk of coronary artery disease. An important complication that arises from this form of causation is that the effects of multiple factors may not be independent, but instead the factors may interact in complex and subtle ways (Hilborn and Stearns 1982). I'll use a classic ecological example to illustrate this point. This example is useful because it was possible to take a rigorous experimental approach, unlike studying interactions between various risk factors that might influence cancer in humans.

In the early part of the 1900s, botanists began to wonder what accounted for geographic variation in the size and appearance of plants that grow in different natural environments. For example, a wildflower called yarrow, with feathery, fernlike leaves and round, flat clusters of small white flowers, grows in sunny spots at a wide range of elevations. Yarrow plants found at high elevations in the Sierra Nevada of California are much shorter than those found in the foothills. Does this variation reflect genetic differences between populations of yarrow at different sites? Or is the variation simply a direct consequence of the effects on plant growth of environmental factors such as length of the growing season, average temperature, available moisture, or soil nutrients?

To answer these questions, Jens Clausen and his colleagues (1948) at the Carnegie Institution of Washington did an extensive series of experiments at various sites in California in the 1940s. They first collected seeds from populations at various elevations and grew them in a common garden at Stanford. Then they selected approximately 30 plants from each of the populations in this garden and divided these plants into three parts, or clones. One clone of each plant was replanted at Stanford, the second clone was planted in a common garden at Mather in the central Sierra Nevada, and the third clone was planted in a common garden at Timberline, near Tioga Pass in Yosemite National Park. Mather is at 1,400 m (4,600 ft) elevation in dense conifer forest; Timberline is at 3,050 m (10,000 ft) elevation near the treeline. Clausen's group followed these plants for 3 years. The most striking result was that

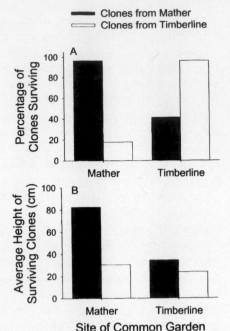

Figure 6.1. Survival (A) and height of the longest stems (B) of clones of yarrow plants (*Achillea lanulosa*) from two source populations at different elevations in the Sierra Nevada of California grown in common gardens at the two sites. Mather is at 1,400 m elevation on the west side of the range; Timberline is at 3,050 m elevation just east of the crest. Data from Clausen et al. (1948).

clones from the Mather population survived much better when planted at Mather than when planted at Timberline, but clones from the Timberline population survived better when planted at Timberline than when planted at Mather (Figure 6.1A). Of the plants that survived, Mather clones were much shorter at Timberline than at Mather, whereas Timberline clones were similar in height at the two sites (Figure 6.1B).

These results suggest that genetic differences between the two sources of plants affected growth because plants from the two sources had different survival probabilities and grew to different heights in the same common garden. But environmental factors also influenced growth because each type of clone responded differently to the different environments of the two common gardens. Most important, however, genetic and environmental effects on growth were not independent because plants from Timberline clones outperformed plants from Mather clones in the Timberline garden but did much worse than plants from Mather clones in the Mather garden (Figure 6.1). This illustrates a *genotype-environment interaction:* the genetic characteristics of Timberline plants that enabled them to do well at high elevation apparently contributed to very poor survival at lower elevations. Genotype-environment interactions will become important in interpreting the story of cancer in Scandinavian twins that is the main focus of this chapter.

There is one more complexity in the concept of causation that is especially relevant for biological phenomena: the idea that causes operate at different hierarchical levels so that a cause at one level is not an alternative to a cause at a different level but is complementary to causes at other levels. I have already hinted at this idea by suggesting that the cause of cancer at the cellular-molecular level is well understood compared to causes at the level of the individual organism, that is, risk factors that make some individuals more likely to get cancer than others. Chapter 7 is dedicated to exploring levels of causation in detail, but it is worth keeping the general idea in mind as you read the rest of this chapter.

GENES AND CANCER: TWO EXAMPLES

So, what *is* the cause of cancer? It should be clear from the last several pages that this question doesn't have a single, simple answer. A lot has been learned about cancer, especially at the cellular and molecular level. Many kinds of cancer can now be detected in early stages, broadening the range of treatment options and improving prognoses. But cancer remains one of the main causes of death, especially in developed countries. What factors increase the likelihood that an individual will experience the sequence of mutations in a cell line in a specific type of tissue that will eventually lead to cancer? In particular, what role do a person's inherited genes (genetic causation) play compared to his or her environment in increasing the risk? In terms of levels of causation, we need to learn more about inherited and environmental factors to complement the knowledge that has been gained through recent research on cellular and molecular factors.[2]

There are several examples of specific genes that are inherited and greatly increase a person's probability of acquiring certain forms of cancer. The best-understood is retinoblastoma, a rare childhood disease in which tumors form in one or both eyes (Murphree 1997). Retinoblastoma occurs about once in every 20,000 to 30,000 live births. The disease is caused by a mutated form of a tumor-suppressor gene called RB, which is located on the thirteenth of the 24 unique types of chromosomes in human cells.[3] The typical form of this gene (RB^+) limits cell division (i.e., suppresses tumor formation); the mutant form does not. Since all cells in an individual except mature sperm and eggs contain two copies of each chromosome, a cell may have two copies of the typical form of the RB gene (symbolized RB^+RB^+), one copy of the typical form and one copy of the mutated form (RB^+RB^-), or two copies of the mutated form (RB^-RB^-). One copy of RB^+ is sufficient to regulate cell division, so only an RB^-RB^- cell is likely to form a tumor. A child may inherit RB^+ from each parent, but in the process of the eye's development, one retinal cell may experience two spontaneous mutations that change each copy of RB^+ to RB^-, producing a tumor that arises from the doubly mutated cell. Because the mutation rate is very low, about one in 1 million to one in 10 million cell divisions, the frequency of so-called sporadic cases arising from these somatic mutations is also relatively low, about one in 50,000 live births.[4] This process invariably produces a tumor in only one eye, or unilateral retinoblastoma.

A child may also inherit an RB^- gene from one parent. The parent may have suffered from retinoblastoma or simply may be a healthy carrier of the RB^- gene. Alternatively, there may have been a mutation converting RB^+ to

RB^- in the production of sperm or eggs in the parent's gonads (in most cases, this occurs in the production of sperm by the father, and the likelihood increases with his age). This child can be symbolized by RB^+RB^-, indicating the types of RB genes that occur on the two copies of chromosome 13 in all the cells of his or her body. In this case, only one additional mutation is required in a retinal cell to produce retinoblastoma. Therefore, a child who inherits one RB^- gene has about a 90% chance of developing retinoblastoma, and two-thirds of the time tumors will develop in both eyes.[5] If the parent suffered from retinoblastoma or is a carrier, each child has about a 45% probability of getting the disease (the chance of inheriting the gene times the chance of developing retinoblastoma if the gene is inherited equals 50% × 90% = 45%). On the other hand, if there was a mutation in the sperm or egg that produced the child with retinoblastoma, the probability that siblings of this child will be afflicted is very low; however, the probability that offspring of the child will be afflicted is 45% for each offspring (Draper et al. 1992).

Fortunately, if diagnosed early, retinoblastoma can be treated successfully so that patients may keep their vision, live normal lifespans, and have children. Effective treatments that preserve vision and minimize the chance of metastasis of the tumor to other organs are relatively recent, however. In the past, there would have been strong natural selection against the RB^- gene because few people with retinoblastoma would have survived to reproduce; most cases of the disease probably resulted from new mutations. Medical advances have weakened this selection against RB^-, leading to the prediction by the human geneticists Vogel and Motulsky (1997) that the overall incidence of the disease might increase from one in 25,000 to one in 10,000 live births over the next 10 generations. In fact, the World Health Organization has reported that retinoblastoma has increased in prevalence over the past few decades (Murphree 1997), consistent with the prediction of Vogel and Motulsky.

Breast cancer is another form of cancer for which there are specific genes that greatly increase risk (Newman et al. 1997; Steel 1997). This is the most common form of cancer in women, each woman having about an 11% probability of developing breast cancer at some time during her life. The primary genes associated with increased risk are $BRCA1$ and $BRCA2$ (BReast CAncer 1 and 2). Women who inherit one copy of a mutant form of $BRCA1$ or $BRCA2$ have about a 70% chance of developing breast cancer by age 70; these mutant forms of the $BRCA$ genes also increase the risk of ovarian cancer in women, as well as prostate cancer and breast cancer in men. As with retinoblastoma, a second, somatic mutation in the breast tissue itself is required for a carrier of a mutant form of $BRCA1$ or $BRCA2$ to develop breast cancer. The typical forms of these genes presumably function as tumor suppressors, just like RB^+, although the molecular details aren't as clear for breast cancer as they are for retinoblastoma. Although the inheritance of these two genes has been well established through the analysis of pedigrees, so that risks to sisters and daughters of people who carry mutant forms of $BRCA1$ or $BRCA2$ can be accurately predicted, only about 5 to 10% of breast cancer cases can be attributed to these specific genes. Therefore, the cause of this common form of cancer is unknown in most cases.

Mutant forms of *RB*, *BRCA1*, and *BRCA2* have large effects, in the sense that they greatly increase susceptibility to particular types of cancer. Because of these large effects, scientists have been able to discover the chromosomal locations of the genes, isolate the genes and determine their DNA sequences, and learn something about their functions. However, many other genes may individually make small contributions to cancer risk. In these cases, the detailed studies of family pedigrees that have been fundamental for developing an understanding of genes with large effects on cancer aren't useful because few cases are likely to show up in any individual pedigree. Instead, large-scale surveys of twins such as the one by Lichtenstein and his colleagues (2000), may reveal the magnitude of the cancer risk due to genes with small individual effects, although not their mechanisms.

CANCER IN TWINS

I began this chapter by describing the different spins on the Lichtenstein study given by different newspapers, with the *Washington Post* emphasizing environmental contributions to cancer risk and the *New York Times* emphasizing genetic contributions. What did Lichtenstein's group actually say? Here are the conclusions of their article in the *New England Journal of Medicine:* "Inherited genetic factors make a minor contribution to susceptibility to most types of neoplasms. This finding indicates that the environment has the principal role in causing sporadic cancer. The relatively large effect of heritability in cancer at a few sites (such as prostate and colorectal cancer) suggests major gaps in our knowledge of the genetics of cancer" (2000:78). This seems more consistent with the *Washington Post* interpretation of the study than with the *New York Times* version, except that Robert Hoover of the National Cancer Institute summarized the work in an editorial in the same issue of the *New England Journal of Medicine* as follows: "Although environmental effects may predominate, the findings with regard to heritability are noteworthy . . . estimates of the proportion of susceptibility to cancer that was due to heritable effects ranged from 26 percent to 42 percent for cancer at the five common sites" (2000:135). To understand these different interpretations of the twin study of Lichtenstein and his colleagues, we need to look at the actual data.

To review, Paul Lichtenstein and eight coauthors from Sweden, Denmark, and Finland analyzed data for about 45,000 pairs of twins. About 35% of these were monozygotic, thus genetically identical, and 65% were dizygotic, sharing 50% of their genes on average. The total incidence of all forms of cancer was about 12%. Both twins had cancer in 1,291 pairs, one or the other had cancer in 8,221 pairs, and neither had cancer in 35,276 pairs. Assuming that the simple fact of being a twin doesn't increase or decrease the risk of cancer compared to people who aren't twins, we can use these data to compute a standard statistic of epidemiology called *relative risk*. Let's use breast

Table 6.1. Breast cancer in female twins in Sweden, Denmark, and Finland as reported by Lichtenstein and colleagues (2000).

	Twin Has Had Cancer	Twin Has Not Had Cancer
Monozygotic Twins		
Focal Individual Has Cancer	84	505
Focal Individual Does Not Have Cancer	505	15,780
Totals for Monozygotic Twins	589	16,285
Probability of Breast Cancer	84/589 = 0.143	505/16,285 = 0.031
Dizygotic Twins		
Focal Individual Has Cancer	104	1,023
Focal Individual Does Not Have Cancer	1,023	28,552
Totals for Dizygotic Twins	1,127	29,575
Probability of Breast Cancer	104/1,127 = 0.092	1,023/29,575 = 0.035

cancer in women who are monozygotic twins to illustrate this calculation. The relative risk of getting breast cancer for a woman in this situation is her probability of getting breast cancer if her twin has already been diagnosed with breast cancer divided by her probability of getting breast cancer if her twin has not been so diagnosed. These probabilities are 0.143 and 0.031, respectively, so relative risk is 0.143/0.031 = 4.6 (Table 6.1). This means that a woman's chance of getting breast cancer is 4.6 times as great if she has a monozygotic twin with breast cancer than if she has a monozygotic twin without breast cancer.

Let's pursue this example a little further. First, what is the relative risk of getting breast cancer for dizygotic twins of women with breast cancer compared to women who have monozygotic twins with cancer? For the former, the relative risk is 0.092/0.035 = 2.6, considerably less than the value of 4.6 for monozygotic twins (Table 6.1).[6] The greater relative risk for individuals who share 100% of their genes with a breast cancer victim than for individuals who share only 50% of their genes with a breast cancer victim implies that genetic factors contribute to breast cancer. However, in a detailed molecular study of a small subset of their monozygotic twins, Lichtenstein and his colleagues found that only two of 12 pairs who both had breast cancer carried mutant forms of *BRCA1* or *BRCA2*, specific genes known to cause cancer. Therefore, there must be other unknown genes that also influence the risk of acquiring breast cancer.

Second, how do relative risks associated with genetic factors compare to relative risks for nongenetic factors? Here are a couple of examples. The relative risk for a woman who experienced menarche (her first menstrual period) before age 12 is about 1.3 times that of a woman who experienced menarche after age 15. The relative risk for a woman who had her first pregnancy after age 24 is about 1.6 times that of a woman who had her first pregnancy before age 20. The relative risk for a woman who was never pregnant is about 1.9

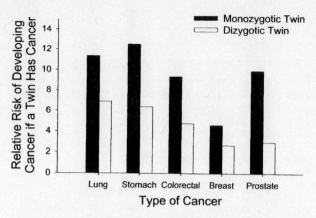

Figure 6.2. Relative risks of developing five common types of cancer. Relative risk is the probability of developing cancer if the twin has cancer divided by the probability of developing cancer if the twin does not have cancer. Data from Lichtenstein et al. (2000).

times that of one who was pregnant before age 20. These and related values for risk are part of a general pattern, suggesting that women who experience more menstrual cycles during their lives (early menarche, late menopause, and few pregnancies) are somewhat more likely to get breast cancer than those who experience fewer cycles. Finally, the relative risk of breast cancer for women who have one or more alcoholic drinks per day is 1.4 times that for women who don't drink (Steel 1997).[7]

In the Scandinavian twin study, there were only five of 28 types of cancer for which there were more than 10 pairs in which both twins were affected: lung cancer, stomach cancer, colorectal cancer, breast cancer in women, and prostate cancer in men. For all five of these common forms of cancer, relative risks were higher if someone had an afflicted monozygotic twin than if someone had an afflicted dizygotic twin (Figure 6.2). These results provide further hints that genetic factors contribute to increased risk of cancer, but they are unsatisfying in at least two respects.

First, twins share not only genes but also common environments to a greater extent than unrelated individuals. The fact that a person's relative risk of developing cancer is substantially greater than one if the person has a twin with cancer could be due to shared genes or shared environmental factors between the two individuals, although the greater relative risks for monozygotic twins than for dizygotic twins implicate genetic factors more directly. Second, and more generally, this analysis doesn't compare the risks due to genetic factors and environmental factors, for which a more detailed analysis of the data is necessary. In particular, I will describe how a quantitative model of the situation can help us think about the relative contributions of genetic *and* environmental factors to the causation of cancer.

ANALYZING GENETIC AND ENVIRONMENTAL CONTRIBUTIONS TO CANCER RISK

In developing their model, Lichtenstein's group imagined that there are many genes that may influence the risk of developing a specific form of cancer, each with a small individual effect on total risk. This is a standard approach in a field called *quantitative genetics*. The approach has been very successful in helping to predict the effects of selection for increased milk production by cattle or egg size of chickens, not to mention selection for bristle number and other characteristics of fruit flies in laboratory experiments. A classical human example of this approach is analysis of genetic contributions to height. Tall parents typically have taller than average children and vice versa; that is, the height of parents and offspring is correlated. This suggests some genetic basis for stature in humans, although environmental factors such as nutrition certainly play a role as well. No one has identified a specific gene that makes a person tall or short. Instead, geneticists assume that many genes influence height, each with a small individual effect. If A and a are alternative forms of one gene, B and b are alternative forms of another gene, and so on, and if the forms indicated by capital letters tend to promote more growth in height than the forms indicated by lowercase letters, a person carrying the genes AABBCCDDEEFFGG is likely to be taller than a person with AabbCCddeeFfGG, who is likely to be taller than a person with aabbccddeeffgg. The remarkable thing about this type of analysis is that it can help us estimate the relative contributions of genetic and environmental factors to variation in a trait such as height in a population, even though we have no idea about what specific genes influence height, what their physiological effects are, or even how many genes are involved.

Think about a well-defined population of individuals, such as adult males of Swedish ancestry living in Sweden. These individuals will vary in height, perhaps from about 1.5 meters (5 feet) to 2.1 meters (7 feet). This variation is called *phenotypic variance* because it refers to an aspect of physical appearance, or phenotype. Swedish males also vary in their genotypes, or genetic composition, as illustrated above. This variation with respect to height is called *genetic variance*. In addition, while they were growing up, the members of the population will have encountered a host of different environmental factors that influenced their eventual height. The variation in these factors is called *environmental variance*, and in principle the total phenotypic variance can be partitioned into genetic and environmental components:

Phenotypic variance = genetic variance + environmental variance.

These components of phenotypic variation can be subdivided in various ways. For studying similarity of traits in twins, it is useful to divide the environmental variance into shared and nonshared environmental factors. For example, twins share the common environment of their mother's uterus during gestation, as well as environmental factors they experience while growing up together in the same household, such as similar exposure to secondhand smoke if one of their parents is a smoker. They also experience a large num-

ber of different environmental factors, ranging from the fact that they develop in different positions in the uterus, so even their gestational environments aren't completely identical; to the fact that they are treated differently
by parents, siblings, and others as children; to the many differences that
occur when they go their own ways as adults (these later differences presumably don't influence height but may influence susceptibility to cancer). Therefore, we can rewrite the equation for phenotypic variance as

$$V_P = V_G + V_C + V_E,$$

where V_P = phenotypic variance, V_G = genetic variance, V_C = shared or common environmental variance, and V_E = nonshared environmental variance.

The key idea from quantitative genetics used by Lichtenstein's group to
estimate the relative contributions of genetic and environmental factors to
susceptibility to cancer is that these components of phenotypic variability are
related to correlations between relatives in their phenotypic values. I introduced correlation analysis in Chapter 2 in describing relationships between
antioxidant levels in blood and memory ability (see Figure 2.1). In the case of
cancer, the data in Table 6.1 can be expressed as correlations in the risk of
getting breast cancer between monozygotic twins or dizygotic twins. These
correlations are 0.366 and 0.255. In one respect, these values are simply one
way of expressing the greater similarity in risk of developing cancer for
monozygotic twins than for dizygotic twins. As we've already seen, another
way to show this is to compare relative risks, which are 4.6 for monozygotic
twins and 2.6 for dizygotic twins. The advantage of using relative risks is that
these figures are easy to interpret as the increase in the probability of getting
cancer if your twin has cancer compared to the probability if your twin
doesn't have cancer. The advantage of using correlations is that relative contributions of genetic and environmental factors to risk can be estimated, as
long as we keep in mind the assumptions that underlie the calculations.

Appendix 1 shows how genetic and environmental contributions to variation among individuals in cancer risk can be estimated from correlations in
risk between monozygotic and dizygotic twins. Using this method, Lichtenstein's group estimated that 27% of the variation in the risk of breast cancer
among female Scandinavian twins was due to genetic factors, 6% was due to
shared environmental factors, and 67% was due to environmental factors
unique to each individual. Figure 6.3 shows that estimates of these components for the five common forms of cancer were fairly similar. The estimated
contributions of genetic factors ranged from 26% for lung cancer to 42% for
prostate cancer, the estimates for shared environmental factors were all generally small, and the estimates for nonshared environmental factors were between 58% and 67%. Thus Lichtenstein's group concluded, "Inherited genetic factors make a minor contribution to susceptibility to most types of
neoplasms. This finding indicates that the environment has the principal role
in causing sporadic cancer" (2000:78). However, the results also justify
Hoover's editorial summary in the *New England Journal of Medicine* that estimates of the role of genetic factors in causing common forms of cancer were

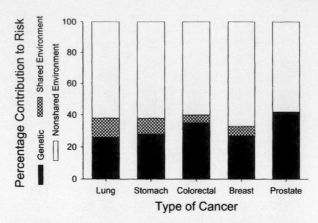

Figure 6.3. Estimated percentage contributions of genetic factors, environmental factors shared between twins, and environmental factors unique to individual twins to the risk of acquiring cancer for Scandinavian twins. Data from Lichtenstein et al. (2000).

"noteworthy." The genetic contributions were all substantially less than those due to nonshared environmental factors, but they were significantly greater than zero for colorectal cancer, breast cancer in women, and prostate cancer in men.

ALTERNATIVE PERSPECTIVES ON CAUSES OF CANCER

This Scandinavian twin study got immediate attention in the popular press but no follow-up stories, although subsequent scientific contributions criticized the interpretations of Lichtenstein's group on various grounds. The critiques are interesting because they illustrate some fundamental points about the analysis of causation, which is the underlying theme of this chapter. Several commentators disagreed with the central conclusion of Lichtenstein's group that environmental factors are of primary importance in causing common forms of cancer. The arguments of these critics were intended to undermine not only this specific conclusion of the Scandinavian twin study but also the widespread opinion among cancer researchers that environmental factors are much more significant than genetic factors in causing cancer. For example, in a 1996 review, the Harvard Center for Cancer Research attributed only 5% of cancer deaths in the United States to hereditary factors. By contrast, 30% of cancer deaths were attributed to tobacco and 30% to dietary factors and obesity, with smaller percentages for a variety of other environmental factors. This group concluded that basic changes in lifestyle might prevent more than half of cancer deaths in the United States (Colditz et al. 1996, 1997).

The belief that cancer is caused primarily by environmental factors is based on several kinds of evidence besides family studies like that of Lichtenstein and his colleagues. First, there is a huge amount of comparative data on incidences of particular forms of cancer in people exposed to various environmental factors. The classic example is, of course, the much greater incidence of lung cancer and many other kinds of cancer in smokers than in non-smokers, but there are many other examples. Eating excessive amounts of saturated fat increases the risk of colorectal cancer. Frequent sunburns, especially during childhood, increase the risk of skin cancer later in life. Exposure to toxic chemicals can cause various forms of cancer.

A second kind of evidence for the importance of environmental factors is that rates of specific forms of cancer in ethnic populations typically approach rates in host countries in a few generations, even though rates in countries of origin may have been quite different. For example, the risk of breast cancer for third-generation Japanese women in the United States is comparable to that for Caucasian women in the United States, but about four times as great as the risk for Japanese women in Japan. Three generations is too little time for significant changes in the frequencies of genes that influence susceptibility to breast cancer, implying that environmental differences between the United States and Japan must account for most of this large difference in risk.[8]

In contrast to the traditional belief that cancer is caused primarily by environmental factors, the recent success of the Human Genome Project has spurred enthusiasm for finding specific genes that contribute to cancer risk. The search has been facilitated by technological advances associated with the project and with progress in molecular biology more generally. This search has had several successes, such as identification of the genes *BRCA1* and *BRCA2*, which have major effects on the risk of breast cancer. The search has not been uniformly successful, however. For example, *BRCA1* and *BRCA2* account for at most 10% of the cases of breast cancer, and two genes recently implicated in prostate cancer account for a similarly low percentage of cases of this disease. Thus there are two competing themes in contemporary research on causes of cancer: one that emphasizes environmental factors and the preventability of many forms of cancer, another that emphasizes genetic factors and development of treatments rooted in a detailed understanding of the genetics and molecular biology of the disease. Does the Scandinavian twin study provide strong support for the environmental perspective in this general controversy, or are the conclusions of Lichtenstein's group compromised so much by their assumptions that the study doesn't help resolve the conflict?

LIMITATIONS OF THE SCANDINAVIAN TWIN STUDY

Several aspects of any scientific study should be evaluated to assess its credibility. For an observational study, we need to consider potential limitations in the data, as well as problems with the models and assumptions used to draw conclusions. I'll focus mainly on the latter, but I'll comment briefly on the former with respect to the Scandinavian twin study. You already know that this was by far the largest study to date of cancer incidence in twins. Despite the impressive sample size, however, there were only enough data to estimate genetic and environmental contributions to risk with confidence for five common types of cancer. There is no good reason to doubt the accuracy of the data themselves in terms of correct identification of twins and diagnoses of cancer. However, one potential problem was pointed out by the geneticist Neil Risch (2001): many of the twins were monitored for a limited period of time and were not particularly old when the data were tallied for analysis. For example, about 29% of the twins were born in Sweden between 1926 and 1958, were still alive in 1972, and were followed from 1973 through 1995. This is only 22 years in which to be diagnosed with cancer, and the twins born in 1958 were only 37 years old in 1995. Many of these twins might still get cancer. In fact, the total incidence of cancer in this group was only 4.5%. By contrast, the Danish twins were born between 1870 and 1930 and followed from 1943 through 1993. The youngest of these twins was 63 in 1993. Therefore there was a much greater opportunity for cancer to develop in this group than in the Swedish group, and the total incidence of cancer in the Danish sample was 21%. Almost 60% of the complete sample from Sweden, Denmark, and Finland consisted of twins who had not yet been exposed to anything near their total lifetime risk of acquiring cancer. The additional cases of cancer that develop in these twins in future years could change estimates of genetic and environmental contributions to risk in either direction, depending on how many of the cases occur in both members of monozygotic pairs, both members of dizygotic pairs, and single members of the two types of pairs. As the study continues and more of these twins are diagnosed with cancer, Lichtenstein's group will be able to make increasingly accurate estimates of lifetime risks and therefore of genetic and environmental contributions to risk. However, Risch argues convincingly that the present data are inconclusive—which illustrates how difficult it can be to do meaningful large-scale, long-term studies. Almost 45,000 pairs of twins is a large number to keep track of but not enough to say much about most forms of cancer; 22 years is a long time to keep track of them but not long enough to thoroughly study a disease that can take a very long time to develop.

Risch's second main criticism of the study by Lichtenstein's group was more fundamental, as well as more abstract. He argued that the basic model used by the Scandinavian group to analyze their data was inappropriate. This model assumes that many different genes have small additive effects on susceptibility to each type of cancer. The combination of genes carried by an individual determines his or her position on a scale of liability or risk. Liability cannot be directly measured, but the model assumes that there is a threshold of liability such that a value above it leads to a specific type of cancer and a value below it does not. An alternative model favored by Risch is that susceptibility is determined by a small number of genes that may have large individual effects but are relatively uncommon in human populations, so that one person is very unlikely to carry more than one of these genes. Unfortunately, the data from the Scandinavian twin study are inadequate to distinguish between these two models. Nevertheless, Risch's analysis illustrates that conclusions in science arise from an interplay between theory and data. Quite different conclusions may result from different models applied to the same data. If Risch's model is valid, for example, the Scandinavian twin data might implicate a larger role of genetic factors in causing cancer than suggested by Lichtenstein's group. In general, Risch's approach is rooted in the research program that focuses on hunting for specific identifiable genes that contribute to cancer. Indeed, he begins his article with a tribute to the Human Genome Project: "Last June, human genome scientists announced completion of . . . a rough draft of the human genome sequence, ushering in a new era of human molecular genetics. This accomplishment was heralded with great fanfare and with predictions of a significant impact on the understanding and treatment of . . . cancer" (Risch 2001:733).

GENETIC AND ENVIRONMENTAL FACTORS AS INTERACTING CAUSES OF CANCER

Let's assume that the model used by Lichtenstein and his colleagues as a foundation for their analysis is valid. Are there any problems with their application of this model? I described several assumptions of the model when I initially explained it, but I didn't discuss the most important assumption—that genetic factors and environmental factors that influence risk are independent of each other; that is, there is no interaction between a person's environment and genotype that affects the likelihood of developing cancer. Without this assumption, it would have been impossible to estimate the contributions of genetic factors, environmental factors shared between twins, and nonshared environmental factors. But this is a big assumption, so we should think about whether it's reasonable or not. We know from the yarrow example earlier in the chapter (Figure 6.1) that genetic and environmental factors *may* interact in determining the phenotype of an organism: yarrow plants adapted to high elevations survived better than plants adapted to low elevations in a common garden at high elevation, but the opposite occurred in a common garden at low elevation. Is there any evidence for a similar kind of interaction for cancer?

In fact, cancer researchers have devoted a good deal of attention to gene-environment interactions. One way in which these interactions might work is that individuals could differ genetically in their ability to metabolize particular carcinogens. For example, various forms of genes code for enzymes involved in the breakdown of compounds in tobacco, alcohol, and food, as well as organic solvents used in industrial processes. Some forms of these enzymes are more effective than others, suggesting that there is genetic variation for the ability to detoxify carcinogens in these substances. This means that two individuals might have similar histories of smoking but different risks of developing lung cancer, depending on whether or not they carried genes that coded for effective detoxification enzymes. Ataxia telangiectasia is an example in which researchers have identified specific genes that increase susceptibility to a particular environmental factor (Swift et al. 1991). This is a recessive disease, meaning that an individual has to have two copies of the rare, defective form of the gene to be affected. The disease involves serious neurological defects, heightened sensitivity to ionizing radiation, and risks of developing leukemia and other forms of cancer that are 100 times as great as risks in the general population. A person with one copy of the defective form and one copy of the normal form of the gene has no overt symptoms of disease but has a somewhat elevated risk of developing cancer (e.g., a relative risk of 3.9 for breast cancer). Furthermore, this risk appears to depend on exposure to radiation. What causes cancer in this case, carrying the ataxia telangiectasia gene or being exposed to radiation? The data suggest that neither factor alone increases risk, but the two acting together can do so; that is, risk depends on a gene-environment interaction. To generalize from these examples, it has been estimated that gene-environment interactions may be involved in 80% of cases of the common forms of cancer (Linet 2000; see also Willett 2002).

If gene-environment interactions are important in many types of cancer, the percentages calculated by Lichtenstein's group for risks due to genetic factors, environmental factors shared between twins, and nonshared environmental factors are nonsensical. Their model without gene-environment interactions forces these percentages to add up to 100%, but if gene-environment interactions exist, the sum may be much greater. In this case, estimating relative contributions of genetic versus environmental factors is meaningless. Richard Peto illustrated this point with a simple example in a letter to the *New England Journal of Medicine* in response to the article by the Scandinavian researchers (Peto 2000). Everyone knows that AIDS is caused by a virus called HIV. But imagine that some subset of the human population has an altered form of a gene that codes for the receptor on the surface of cells that is used by HIV to gain entry into the cells. These people would be immune to HIV because the virus particles wouldn't recognize their altered receptor molecules and therefore wouldn't be able to enter their cells and become integrated into their nuclear DNA, causing AIDS. In this scenario, we could say that 100% of AIDS cases are due to an environmental factor, HIV, but also that 100% of cases are due to a genetic factor, having the gene that codes for the typical form of receptor molecules on the surfaces of cells that can be used by HIV. This situation represents a case in which two factors, one environmental and one genetic, are each necessary but not sufficient to cause a disease. This is one of several forms that gene-environment interactions can take. In fact, this example isn't completely hypothetical because Stephen O'Brien and Michael Dean (1997) of the U.S. National Cancer Institute have recently found a mutant form of a human gene that provides immunity to HIV. This altered gene may have evolved in northern Europe, starting about 700 years ago, in response to the epidemics of plague that swept through Europe between 1350 and 1670 (the bacterium that causes plague infects the same cells of the immune system as HIV).

There are various ways of studying gene-environment interactions, ranging from laboratory experiments with animal model systems to detailed epidemiological investigations of human populations. Neither of these basic approaches is completely satisfying, but progress may come from integrating evidence from the two kinds of studies. In theory, a well-designed experiment is a much stronger way to test an hypothesis in which two kinds of factors interact to produce an effect than is a study based on the kinds of data that epidemiologists frequently collect for human populations. In an experiment, genetic and environmental factors can be carefully controlled so that effects can be clearly attributed to specific genetic or environmental differences that exist between the groups being compared. The yarrow example is an illustration: genetic factors were controlled by using genetically identical clones of plants, and environmental factors were controlled by planting different clones in common gardens at various elevations. Using a similar approach to study the causation of a specific disease requires an animal model system that faithfully represents the disease as it appears in humans. Mimics of particular human diseases don't always exist in animals, although there is a long history of using animal models to study general physiological processes, and these types of studies have made fundamental contributions to the basic understanding of the biology of disease. Nevertheless, many diseases can't be solved solely by evidence from animal models.

One epidemiological approach to the role of gene-environment interactions is based on comparing incidences of a disease in relatives who are differentially exposed to an environmental risk factor (Yang and Khoury 1997; Andrieu and Goldstein 1998). For example, suppose Lichtenstein's group expanded the study of Scandinavian twins to compare rates of lung cancer in smokers and nonsmokers who had twins with and without lung cancer. In theory, this approach could be used to assess the relative contributions of genetic factors versus the contributions of a specific environmental factor, exposure to carcinogens in tobacco smoke. It could also be used to test for an interaction between smoking and genetic factors. For example, both exposure to tobacco smoke and a specific set of genes might be necessary for lung cancer to develop, or either alone might be sufficient to produce lung cancer. The problem with this approach is that relatives, especially twins, have many environmental factors in common in addition to the specific factor under consideration. These additional factors make for messy analyses and inconclusive interpretations of twin studies.

A somewhat cleaner approach is possible with epidemiological data if specific genes have been identified that contribute to disease risk. In this case, a large random sample of unrelated individuals can be divided into four categories: those who carry the genes and have been exposed to a particular environmental risk factor, carriers who have not been exposed, noncarriers who have been exposed, and noncarriers who have not been exposed. If the incidence of the disease is higher (or lower) in individuals with both genetic and environmental risk factors than would be predicted by incidences in individuals with only one of these risk factors, a gene-environment interaction is implicated. This approach is limited by the fact that the subjects may differ in many other ways, both genetically and environmentally, that influence susceptibility to the disease; these other factors are uncontrolled and in many cases unknown, so interpretations of the results in terms of the known risk factors may be challenged. However, this approach is about the best that can be done without doing a true experiment, as in the yarrow study, and the experimental approach may be unethical or impractical in the case of human disease.

Lisa Gannett (1999) provided an interesting perspective on the analysis of causation in an article based on her doctoral dissertation in the philosophy of science at the University of Minnesota. She pointed out that all characteristics of organisms are influenced by both genetic and environmental factors and that gene-environment interactions are pervasive if not universal. She then suggested that the reason for current emphasis on genetic causation of human diseases is primarily pragmatic rather than theoretical: "When the pragmatic dimensions of genetic explanations are recognized, we come to understand the current phenomenon of 'geneticization' to be a reflection of increased technological capacities to manipulate genes in the laboratory, and potentially the clinic, rather than theoretical progress in understanding how diseases and other traits arise" (1999:349). According to Gannett, a focus on genetic factors may arise not only from these technological advances but also from social, political, and economic factors that make genetic approaches to disease seem easier or more palatable than attempts to modify environmental factors: "Genetically engineered solutions make private investors money; serious attempts to counter poverty, environmental degradation, and tobacco, alcohol, and drug addiction just cost taxpayers money" (1999:370).

In this chapter, we have dissected a recent study of cancer in twins that was interpreted in contrasting ways in the popular press. We found that the *Washington Post* headline, "Cancer Study Downplays Role of Genes," accurately represents the main conclusion of the authors of the original research. However, to reach this conclusion, the authors had to ignore possible gene-environment interactions, even though there is quite a bit of evidence that these kinds of interactions may be important in causing cancer. We saw that there are different perspectives about the question of what causes cancer. At the molecular level, the cause is genetic, in the sense that cancer depends on mutations in DNA that disrupt the normal regulation of cell growth and division. At the population level, one school of thought is that environmental factors predominate in causing cancer, implying that much cancer is preventable by removing or correcting these factors. Another school of thought emphasizes inherited contributions to risk, partly because of powerful new techniques of molecular biology for identifying genes in family pedigrees that are correlated with particular forms of cancer. Yet a third school, which partly overlaps the first two, focuses on gene-environment interactions. This third approach seems to me to be the most promising because it embraces the complexity of cancer as a scientific challenge rather than trying to oversimplify the problems for the sake of tractability.

I've used this example, not only because cancer is a disease that touches almost all people directly or indirectly and because twins are fascinating subjects for study, but also to illustrate some fundamental general ideas about the analysis of causation in biology. Biological phenomena rarely have simple, straightforward causes. Causes can be recognized at different levels, including external environmental stimuli, genetic differences between individuals, and cellular and molecular processes. Causal explanations at these various levels are complementary and interdependent, and there may be various kinds of interactions at the same or different levels that produce complicated patterns of consequences. Finally and most important, it is these complexities that make science both challenging and rewarding.

RECOMMENDED READINGS

Gannett, L. 1999. What's in a cause?: The pragmatic dimensions of genetic explanations. *Biology & Philosophy* 14:349–374. This is a clear and trenchant review of the meaning of causation illustrated by contemporary human genetic examples.

Hilborn, R., and S. C. Stearns. 1982. On inference in ecology and evolutionary biology: The problem of multiple causes. *Acta Biotheoretica* 31:145–164. If you can locate this article, it will help clarify and illustrate the manifold ways in which causation can be complex.

Trichopoulos, D., F. P. Li, and D. J. Hunter. 1996. What causes cancer? *Scientific American* 275(3):80–87. This issue of *Scientific American* was devoted to cancer, and this article gives a general perspective on its causes.

Willett, W. C., G. A. Colditz, and N. E. Mueller. 1996. Strategies for minimizing cancer risk. *Scientific American* 275(3):88–95. This provides some useful practical advice for reducing cancer risk in individuals and populations.

Chapter 7

Why Do We Age?
Different Levels of Causation as Complementary Explanations

In January 2002, a team of biologists in Lawrence Donehower's laboratory at the Baylor College of Medicine in Waco, Texas, reported a surprising chance discovery in their colony of mice used as a model system to study cellular and molecular processes of cancer. A protein called p53 is a major focus of their work. This is a tumor-suppressor protein, which can forestall cancer by blocking cell division or causing the death of primordial cancer cells.[1] Most of the work in Donehower's lab had involved creating mutations in the gene for p53 that caused reduced activity of this protein and increased the frequency of tumors in mice; but in 1994 the researchers accidentally made a new type of mutation that caused *increased* activity of the p53 protein and *reduced* incidence of tumors, from about 48% in normal mice to about 6% in mice carrying this new mutation.

The researchers initially thought that not much could be learned from studying these mutant mice. If tumors were rare, there would be little opportunity to study their cellular and molecular characteristics to learn more about how p53 works. The biologists set the mice aside and turned to other projects, but about a year later they noticed that the mice looked prematurely old. Stuart Tyner, a graduate student in the lab, took on the task of documenting the aging process in these mice (Tyner et al. 2002). Normal mice in the lab lived an average of 118 weeks; the mutant mice lived an average of 96 weeks. Furthermore, the mutant mice had a host of symptoms of premature aging: weight loss, thinning hair, osteoporosis, atrophy of various internal organs, and delayed wound healing. Although the altered form of p53 protected these mice from cancer, it apparently caused them to age faster and die sooner. These results are fascinating because we normally think of cancer as a common correlate of aging, but in the mutant mice just the opposite occurred.

What are the larger implications of this discovery? One interpretation of the results is that the activity of the p53 protein may be precisely regulated under normal conditions. The fundamental action of p53 is to control the life cycle of cells: how often and how rapidly they divide before they die. This enables p53 to stop the indiscriminate cell division that leads to malignant tumors, but it also means that p53 may eventually stop the division of stem cells necessary for the maintenance of tissues and organs in the body.[2] Too little p53 may lead to an increased probability of getting cancer; too much p53 may lead to premature aging. As expressed by the science journalist Nicholas Wade, "We age in order to live longer" (*New York Times*, 8 January 2002, Section F, p. 1). Without p53, we would probably die early from cancer (as did mutant mice that lacked p53 in the Donehower lab); with p53, we live longer without cancer but eventually show the many signs of aging that are so familiar.

But this hypothesis about how p53 works raises another intriguing set of questions. If the activity of p53 is precisely regulated between the Scylla of cancer and the Charybdis of aging, why do mice typically live for 2 to 3 years and humans for 70 or more? What is the optimal activity of p53 for a species? Do most individuals of a species have nearly optimal amounts? If not, what kinds of constraints influence these amounts? If so, why do species differ in their optimal amounts? These kinds of questions illustrate a different way of thinking about causation than the mechanistic questions asked by members of the Donehower lab. They are questions about the evolution of p53, and more generally about the evolution of aging.

LEVELS OF CAUSATION IN BIOLOGY

These different kinds of questions about cancer and aging illustrate a fundamental principle in biology: causes operate at different levels (ranging from biochemical interactions in cells to evolutionary processes that occur over many generations), and explanations at these different levels are complementary rather than competitive. This principle was first articulated by the evolutionary biologist Ernst Mayr (1961) and the ethologist Niko Tinbergen (1963), although it has often been ignored or misunderstood by researchers who proposed an explanation for something at one level as a false alternative to another explanation at a different level (Sherman 1988; see also Holekamp and Sherman 1989; Armstrong 1991). Biologists have considered the implications of multiple levels of causation most thoroughly for animal behavior, but the principles apply in all areas of biology. I will briefly outline these principles as they apply to behavior and illustrate them with a specific example from human reproduction before returning to the more general topic of what causes aging. The following hierarchy of levels of causation is modified somewhat from that outlined by Paul Sherman in 1988 to better fit my discussion of aging.

1. Specific environmental stimuli induce a behavior. These stimuli may
    come from the physical environment, other species (e.g., predators
    and prey), and members of the same species. For example, a frog re-
    sponds to movement of a small object such as an insect across its field
    of view by a tongue flick as part of its characteristic feeding response.
2. The internal physiological state of an animal influences its likeli-
    hood of performing a behavior. This physiological state includes
    hormonal and neurological components that may be altered by pre-
    dictable daily and seasonal cycles, as well as by learning. For ex-
    ample, males of territorial bird species typically begin singing in
    early spring in response to increases in circulating testosterone, which
    in turn responds to the increased length of days as the spring breed-
    ing season approaches. This level of causation includes intracellular
    and molecular factors, such as the mutations that influence cell divi-
    sion and growth in cancer, as well as aspects of physiology that in-
    volve multiple cells, tissues, and organs, such as the effects of testos-
    terone on singing by male birds.
3. Ontogeny or development of an organism influences the behavior it
    shows later in life. For example, in many songbirds, there is a critical
    period for song learning in the first few months of life. In their first
    breeding season, male birds sing songs that they heard their fathers
    sing the previous summer. This process of imitation produces di-
    alects in some species, so different populations of the same species
    have consistently different songs. Genetic differences between indi-
    viduals also influence behavior and can be classified with develop-
    mental factors because their effects may be delayed until sexual ma-
    turity or later, in contrast to the first two levels of causation that
    have relatively quick effects.
4. Finally, behaviors are influenced by the evolutionary history of a
    population. Behaviors performed by a subset of a population in one
    generation may increase their reproductive success over other indi-
    viduals in the population. If there are genetic differences between
    individuals that influence their ability to perform the behaviors, the
    genes that underlie more successful behaviors will become more fre-
    quent in the population, as will the more successful behaviors. This
    process is called evolution by *natural selection;* it produces traits that
    are adapted to particular environmental circumstances.

These four levels of causation are not alternative, mutually exclusive causes
of behavior. Instead, a full explanation of behaviors and other traits of organisms requires explanations at all levels. More important, explanations at the
different levels are interdependent. Progress in understanding causation at
one level may be stalled until a breakthrough occurs at another level, opening
new avenues of research at the first level.

A human example may help bring this idea of multiple levels of causation
into better focus. Morning sickness is a common complaint of pregnancy.

Nausea and vomiting are experienced by most women during the first 14 to 16 weeks of pregnancy; these symptoms may be associated with changes in appetite but don't reflect poor nutrition or poor general health of the mother or fetus. What causes morning sickness?

The environmental trigger of morning sickness is clearly ingestion of certain foods (level 1 in the list above). A broader range of foods induces nausea in pregnant women than in men or nonpregnant women. The internal physiological mechanisms that control nausea and vomiting involve regions of the brain that receive messages from touch receptors in the digestive tract (level 2). The profound hormonal changes that occur during pregnancy clearly make pregnant women more sensitive to the neurological processes that actually bring about nausea and vomiting (level 2), although there are no predictable differences in the blood concentrations of hormones between women who experience severe morning sickness and those who experience no symptoms or only mild symptoms. Possibly, genetic differences or differences in early development cause some females to be more sensitive than others to these hormonal changes when they reach adulthood and become pregnant (level 3). Finally, several hypotheses have been proposed to explain the evolutionary cause or adaptive significance of morning sickness (level 4). The most comprehensive is that morning sickness is an adaptation to protect the embryo and mother from the effects of toxins and pathogens in foods. Pregnant women have suppressed immune systems, presumably to reduce the chance of developing an immune response to their own offspring during gestation. Vomiting eliminates toxins that might harm the embryo and pathogens that might harm either the embryo or the mother, with her reduced immunity against infection. Nausea may also bring about a learned aversion to associated foods, providing further protection to the embryo and mother. Sherman and Flaxman (2001, 2002) exhaustively reviewed the evidence for this and other hypotheses. One of their most interesting results was that the likelihood of a miscarriage is substantially greater for women who do not experience morning sickness than for women who do. This suggests a clear benefit of morning sickness in terms of increased reproductive success, which could be favored by natural selection.

WHAT IS AGING?

We can apply the idea of complementary levels of causation to many problems in biology, ranging from specific questions about causes of morning sickness to much more general questions about why humans, other animals, and even plants deteriorate as they get older. Before thinking about aging from this perspective, we need to consider what aging means. Each of us probably has an intuitive understanding of the meaning of this term based on older people we know or observe. In fact, we use common cues of appearance, speech, and movement to classify people we don't know as elderly or not. But a more precise definition of aging, or senescence, will be helpful in trying to understand its causes.

In the most general sense, aging is a change in the characteristics of something over time. Aging does not simply mean getting older: a wine glass in your kitchen cupboard doesn't age. However, imagine a bottle of wine in your wine cellar (many of you will have to imagine the wine cellar, too). The wine does age because chemical reactions occur in the bottle that change the taste of the wine. In this example, aging usually brings improvement: the most expensive wines are often the oldest (Ricklefs and Finch 1995). For humans and other organisms, a somewhat more focused definition given by Steven Austad (1997:6) better represents what we usually think of as aging: "the progressive deterioration of virtually every bodily function over time." In other words, aging is not just any change in how our bodies work as they get older, but change for the worse.

For any species, there are a multitude of specific manifestations of aging. It's easy to think of examples for humans. Visual acuity declines, especially at close range. Hearing ability declines, especially for high frequencies. Muscle strength declines by about 1% per year. Males produce fewer sperm, and more sperm are malformed. Females stop releasing eggs from their ovaries at menopause. Many other functional changes characterize aging in humans and other animals. Likewise, botanists describe senescence in plants. Very old trees of some species grow more slowly, produce fewer fruits, and are less productive in general than younger trees of the same species, although bristlecone pines, which are renowned for great longevity, apparently show no overt signs of senescence even beyond 1,000 years of age (Lanner and Connor 2001).

Deterioration in bodily function causes an increase in mortality rate that can be used as a general indicator of aging in many different kinds of organisms. This statement is more subtle than it may seem. For one thing, you might assume that it is a truism: no plants or animals are immortal, so the mortality rate or probability of dying inevitably increases with age. But this idea is based on interpreting *probability of dying* as a cumulative probability rather than the probability of dying per year or per week or per day, whatever time period is appropriate for the type of organism you are thinking about. The probability for a specified time period could conceivably remain constant rather than increasing as an organism gets older. For example, suppose the probability of dying is 1% per year regardless of how old an animal is. Then the probability of living to be 50 years old would be about 60% and the probability of living to be 100 would be about 37%, but we would not want to attribute this difference to senescence.[3] Instead, there is a smaller probability of living to age 100 than to age 50 in this example simply because more years have elapsed in which the animal has been exposed to a constant probability of death per year.

The example in the above paragraph isn't purely hypothetical. Justin Congdon and others have been studying Blanding's turtles in southeastern Michigan since 1953. Since the ages of turtles can be estimated by counting growth rings on the shells, the researchers had records for turtles born as early as 1933 and still alive at age 66 in 1999. Annual mortality rates were, if

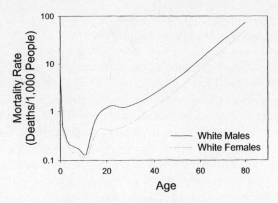

Figure 7.1. Change in annual probability of death with increasing age in white males and females in the United States in the year 1998. Note that the vertical axis is logarithmic—each interval represents multiplication of the annual mortality rate by 10, from 0.1 per 1,000 to 1 per 1,000 to 10 per 1,000 (1%) to 100 per 1,000 (10%). Data from United States Census Bureau (2002).

anything, *lower* for turtles in their 50s and 60s than for turtles younger than 50 (Congdon et al. 2001). Thus senescence may not be universal, whether measured as a decline in performance or as increased probability of dying. Bristlecone pines are just as robust at age 1,000 as at age 500, and Blanding's turtles have at least as good a chance of surviving from age 60 to 61 as from age 30 to 31, although either of these species might show signs of senescence at even older ages.

Figure 7.1 shows how mortality rates change with age in humans. Note that mortality rates for both males and females decline from birth until about age 12, then gradually increase through the rest of the life span. This pattern of a minimum mortality rate at puberty is characteristic of both sexes of every human population that has been studied, although its value differs widely among populations. Many other kinds of animals and plants have mortality rate curves of this shape also. The slope of the increasing part of the graph, which is the rate of increase of mortality rate with age, is a quantitative measure of aging that is useful for comparisons of different human populations or even of different species. This rate can also be expressed as the *mortality-rate-doubling time:* the length of time it takes for each doubling of mortality rate. For white females in the United States, mortality rate doubles about every 8 years from age 25 on. In the year 1998, the probability of dying in the next year was about 0.044% for 25-year-old females. With a mortality-rate-doubling time of 8 years, the annual probability of death would be about 0.09% for 33-year-olds, 0.18% for 41-year-olds, 0.36% for 49-year-olds, 0.72% for 57-year-olds, 1.44% for 63-year-olds, and so on. For white males in the United States, the mortality-rate-doubling time is about 9 years, reflecting the slightly less steep slope of the line for males after age 25. This means that the mortality rate increases slightly more slowly with age for males than for females after age 25, although at all ages through at least age 80 males have a higher mortality rate than females.

If the mortality rate increases with age during some part of the life span, we can use a graph like Figure 7.1 as a quantitative picture of aging. We can also calculate the mortality-rate-doubling time as an index of the rate of aging. This index may differ slightly between different groups, such as white males and females in the United States, or it may differ substantially between different species. For example, mice in laboratory colonies have mortality-rate-doubling times of about 3 months, in contrast to values of about 8 years for most human populations. With aging characterized as an increase in the mortality rate as organisms get older, graphs such as Figure 7.1 also enable us to answer another interesting question: at what age does aging begin? For human beings, the surprising answer is that aging begins at about the time of puberty, when the probability of death in the next year is at a minimum.

One final aspect of the quantitative description of mortality is important for understanding the causes of aging—the idea that sources of mortality can be divided into two categories: extrinsic and intrinsic. Extrinsic sources include things like accidents, infectious diseases, parasitism, predation, and starvation, as well as homicide and suicide for humans. Intrinsic sources include genetically determined diseases and general deterioration in bodily functions with age that are manifested in cancer, cardiovascular disease, Alzheimer's disease, and so on. These two types of mortality are clearly not independent of each other. For example, moose in their prime are largely invulnerable to predation by wolves, but older moose are slower and weaker and can be killed by wolves. In this case, the immediate cause of death might be extrinsic, but the intrinsic condition of the older moose also contributes to their demise. Similarly, the immune systems of humans and other animals deteriorate with age, so older individuals are more vulnerable to infectious diseases than younger ones. Even the impacts of accidents depend on intrinsic factors: an elderly person with osteoporosis who falls off a ladder is more likely to die than a younger person who falls from the same height because the osteoporosis makes it more likely that the older person will sustain serious fractures.

Despite the fact that both extrinsic and intrinsic factors contribute to many deaths in a population, the distinction is useful in interpreting graphs that show how mortality rate changes with age. In Figure 7.1, for example, the different heights of the curves for males and females primarily represent differences in extrinsic mortality. In particular, the hump in the male curve between ages 15 and 25 is due to the greater likelihood of accidental death, homicide, and suicide for males in this age range, when they are afflicted with "testosterone dementia," in Steven Austad's (1997) words. By contrast, the slopes of the steadily increasing portions of the curves beyond age 25 in humans represent changes in the rate of intrinsic mortality. For white males and females in the United States, these lines are roughly parallel (although the line for females is a little steeper than the one for males), which implies that intrinsic mortality increases at about the same rate for males and females. In

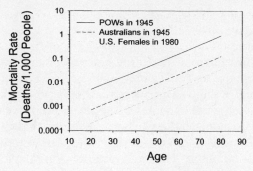

Figure 7.2. Change in annual probability of death with increasing age for three human populations: female Australian prisoners of war held by the Japanese army in World War II, female Australian civilians at the same time, and white females in the United States in 1980. Each line begins near puberty, when the mortality rate is near its minimum. The scale of the vertical axis is logarithmic, as in Figure 7.1. The different positions of the lines represent differences in extrinsic mortality, which was much higher in the Japanese POW camp than in Australia at the same time or in the United States 35 years later. However, the fact that the lines are parallel to each other means that the rate of aging was similar in all three cases. Modified with permission from Figure 1 in "Slow Mortality Rate Accelerations during Aging in Some Animals Approximate That of Humans," by C. E. Finch, M. C. Pike, and M. Witten, *Science* 249:902–905, copyright ©1990 by the American Association for the Advancement of Science.

fact, rates of intrinsic mortality are strikingly similar for human populations in a wide range of situations in which rates of extrinsic mortality differ greatly. For example, Caleb Finch and two colleagues (1990) compared the mortality rate curves of human females in the United States in 1980, Australian females in Australia during World War II, and Australian females in Japanese prisoner-of-war camps on the island of Java in World War II. On a graph showing increasing mortality rate with age, the lines for these three groups were almost perfectly parallel (Figure 7.2), with mortality-rate-doubling times of about 8 years in all three cases, despite the fact that extrinsic mortality rates were 10 to 30 times as great for the prisoners of war as for the other two groups.

PHYSIOLOGICAL EXPLANATIONS OF AGING

With this background, we can now consider the causes of aging, with emphasis on different kinds or levels of causation. As with morning sickness, explanations at different levels are not competing but may instead reinforce particular explanations at other levels. For example, if there is strong evidence in favor of a specific hypothesis about the mechanisms that produce aging, an evolutionary hypothesis that is consistent with this mechanistic explanation would have an automatic advantage over other evolutionary explanations that are not.

Let's begin with physiological causes of aging. If you thought much about aging before reading this book, you probably thought mainly about physiological causes because humans seem to have a natural tendency to wonder how things work (or fail to work, in the case of our aging bodies). You may even be surprised to learn about other kinds of causes later in this chapter that are interesting and important but get less attention in the popular press.

At least 50 physiological hypotheses have been proposed to explain aging (Medvedev 1990). Among all these hypotheses, those that emphasize cellular and molecular mechanisms are the most exciting because they have the potential of broad applicability. In fact, an explanation of aging at the cellular and molecular level might encompass many other physiological hypotheses that involve deterioration of specific tissues or organs or that focus on aging in humans. For this reason and because discovery of broad, sweeping generalizations often leads to fame and sometimes to fortune in science, researchers have devoted most of their attention to studying fundamental cellular and molecular causes of aging. I'll consider two of the major hypotheses at this level. The first hypothesis, that aging is due to inherent limits in how often cells can divide, was proposed in the 1960s and is still popular, although it has some fundamental flaws. The second hypothesis, that aging is due to molecular byproducts of cellular metabolism, is one of the most active topics of research on aging today.

The cell division hypothesis grew out of work by Leonard Hayflick and Paul Moorhead that was published in 1961. By that time, there was already a long history of maintaining certain types of cells in culture dishes containing an appropriate mix of nutrients. Many of these cell cultures could be maintained indefinitely by letting the cells divide until they filled a culture dish, then transferring a few cells to a new culture dish with a fresh nutrient medium. But immortal cell lines like these were derived from cancer cells, which by definition are capable of unrestrained growth. Hayflick and Moorhead, who wondered if normal cells would show the same characteristics in cultures, used fibroblasts, which are cells that form fibrous material in connective tissue.[4] Through very careful techniques, Hayflick and Moorhead consistently found limits to the growth of fibroblast cells in culture. Cells from human fetuses, for example, typically divided about 50 times[5] and then stopped dividing, although they could be kept alive for several years after that.

By itself, this result doesn't say much about whether limited cell division might be an underlying cause of aging because cells in a living body might behave very differently than cells in a culture dish in the laboratory. But Hayflick and Moorhead extended this work in two ways that more directly supported the cell division hypothesis. First, they showed that fibroblasts from adult humans divided only about 20 times in culture before stopping, compared to 50 divisions for fetal fibroblasts. Second, fetal fibroblasts from long-lived species divided more often in culture than those from short-lived species. For example, the "Hayflick Limit" for mouse fibroblasts was 14 to 28 cell divisions in culture, compared to about 50 divisions for human fibro-

blasts and more than 100 for tortoise fibroblasts. The first result was interpreted to mean that limited cell division was not just a trait of fibroblast cells
in culture but also occurred in fibroblast cells in vivo (in life). Lineages of
cells in adults apparently had used up part of their allotment of divisions, so
fewer divisions were possible when they were placed in culture. The second
result was interpreted to mean that species that normally live a long time are
able to do so because they have a mechanism for increasing the number of
times that essential cells can divide.

Further research with cell cultures and living organisms has revealed an
elegant mechanism for the limits to cell division discovered by Hayflick and
Moorhead. The genetic information encoded in DNA is carried in chromosomes in the nuclei of cells. At the ends of these chromosomes are structures
called telomeres that consist of short sequences of DNA bases (the building
blocks of the genetic code) repeated many times. Unlike other sequences of
DNA in chromosomes, the sequences of bases in telomeres do not provide
the instructions for making specific proteins but are part of what is sometimes called junk DNA. However, junk DNA in telomeres has a very important function. Each time the DNA strand in a chromosome is replicated in
cell division, a short piece at the end is clipped off as a necessary consequence
of the mechanical process of replication. This removes a small part of the
telomere. After a certain number of divisions, the entire telomeres of some
chromosomes will be missing. If division continued, functionally important
DNA would eventually be clipped from the bare ends of the chromosomes,
causing the cell to work improperly or die. Instead, the removal of telomeres
from the ends of chromosomes is a signal that causes most normal cells to
stop dividing. There are two exceptions to this pattern. Cancer cells have an
enzyme called telomerase that rebuilds the telomeres and therefore allows
the cells to keep dividing. So do reproductive cells, which belong to cell lineages that persist from generation to generation (Marx 2002).[6]

What's wrong with the idea that limited cell division is a basic mechanism
of aging? Steven Austad wrote that "major medical textbooks often cite the
Hayflick Limit as the fundamental essence of aging" (1997:65), although
most gerontologists, including those who have worked for many years on
limits to cell division, don't believe it. There are two major problems with
the cell division hypothesis. First, many types of cells in our bodies live and
perform normally for many years without dividing; that is, continuous cell
division is not a prerequisite for normal function of all tissues, although it is
for some. For example, muscle cells stop dividing at or shortly after birth in
humans and other mammals. Weight-bearing exercise increases muscle mass
by causing enlargement of individual cells, not division of muscle cells (Austad 1997). Second, limits to cell division are really a necessary part of normal
development, not simply a consequence of the biology of cells that causes
aging and eventual death. Without these limits, cells couldn't form the amazing diversity of sizes and shapes that make up living organisms. Indeed, quite
in contrast to the hypothesis that aging is caused by limits to cell division,
one of the most common signs of aging in humans and many other animals is

cancer, which results when normal limits to cell division are broken. The alternative hypothesis that aging is due to molecular byproducts of cell metabolism helps resolve the paradox that cells in some kinds of tissues must retain the ability to divide to sustain life but that the unlimited cell division of cancer often causes death.

This alternative, mechanistic hypothesis for aging is usually called the free-radical hypothesis. It was first proposed by Denham Harman in 1956 and is based on the fact that all cells contain mitochondria, where energy needed for functioning of the cells is processed. The energy processed by mitochondria is contained in the chemical bonds of glucose, a simple sugar that is derived from several different types of nutrients in an animal's diet. A series of chemical reactions involving oxygen transfers this energy to other compounds that participate in functionally important processes in cells, such as the sliding of long protein molecules that produces the lengthening and shortening of muscle fibers. Almost all of the oxygen used in the mitochondrial chain of reactions is converted to water, but small amounts are converted to byproducts called reactive oxygen species, as described in Chapter 2. These byproducts can damage DNA, proteins, and lipids, which are fundamental constituents of all cells. Many of the most damaging reactive oxygen species are free radicals, molecular fragments that have one or more unpaired electrons; this is the source of the designation free-radical hypothesis.

The free-radical hypothesis offers a relatively straightforward explanation of aging: that the gradual accumulation of damage in *cells* due to reactive oxygen species causes most if not all of the structural and functional deterioration of *bodies* that occurs as organisms grow older. For example, atherosclerosis is a thickening or hardening of the walls of arteries due to plaque deposits in which LDL cholesterol is a major component. Atherosclerosis can lead to heart attack, stroke, and other cardiovascular diseases. Plaque deposits on the inner walls of arteries may begin to form when reactive oxygen species make LDL cholesterol susceptible to attack by the immune system, producing clumps of cells that settle into arterial walls. Likewise, Alzheimer's disease is characterized by distinctive clumps of proteins, called amyloid plaques, in the brain. Their formation is apparently stimulated by reactive oxygen species, and in a positive feedback loop, these plaques induce the formation of more reactive oxygen species, which damage nerve cells. Finally, reactive oxygen species may cause some of the mutations in DNA that lead to cancer (see Chapter 6 for more details). In other words, cancer may be a result of aging because cumulative damage from reactive oxygen species disrupts the controlled cell division necessary for normal life.

Various lines of evidence are consistent with the free-radical hypothesis of aging. Kenneth Beckman and Bruce Ames (1998) of the University of California at Berkeley listed 14 categories of evidence, including age-related changes in constituents of cells that result from oxidative damage, genetic experiments with model species of invertebrates, and comparison of human populations that differ in longevity and amounts of antioxidants in their diets. The most convincing evidence at this time probably comes from genetic studies of roundworms (*Caenorhabditis elegans*) and fruit flies (*Drosophila melanogaster*). These are model organisms for studies of genetics and development because it has been possible to create many mutant forms of each species that show various patterns of growth and development. Researchers can study the mechanisms of development from a fertilized egg to an adult worm or fly by linking these various patterns of growth and development to differences in proteins that are synthesized by the mutant forms.

One mutant of *Caenorhabditis* is called *age-1*. The maximum life span of *age-1* worms is more than twice as long as that of wild-type worms,[7] and their average life span is about 65% longer. In standard tests using hydrogen peroxide as an oxidizing agent, *age-1* are more likely to survive than wild-type worms, and the survival of *age-1* worms exposed to hydrogen peroxide actually increases with age, whereas that of wild-type worms doesn't change. At least two antioxidant compounds that help repair damage in cells from reactive oxygen species are more active in *age-1* worms than in wild-type worms, and activity increases with age in the former. These antioxidant compounds are natural constituents of the cells of many organisms, including humans, so experiments with *Caenorhabditis* may have broad relevance for understanding the mechanisms of aging.

In addition to inducing specific mutations by using radiation or chemical treatments, fruit fly researchers use selective breeding or artificial selection experiments in laboratory bottles to investigate genetic processes of development. Michael Rose (1984) of the University of California at Irvine did such an experiment by using eggs from older and older flies to start each new generation. In each generation, therefore, the oldest reproducing females were selected as mothers of the next generation. Rose wanted to see if the average longevity of flies selected in this way would increase, and indeed it did. After several generations, the average life span of female flies in five selected lines was 43 days compared to 33 days in five control lines, in which eggs were chosen at random to start each new generation. In addition, the frequency of a gene that codes for an active form of an antioxidant called SOD (superoxide dismutase) was between 10 and 28% in the five lines of flies selected for longevity but near 0% in the five control lines. It's probably not coincidental that SOD was one of the antioxidants that was more active in long-lived *Caenorhabditis* than in wild-type worms (Beckman and Ames 1998).

ENVIRONMENTAL CAUSES OF AGING

Physiological causes of aging are diverse, but most may be related to a fundamental molecular mechanism, the damaging effects of reactive oxygen species on DNA, proteins, and lipids. The free-radical hypothesis, however, is far from a complete and final explanation of aging. We might ask, for example, if there are any factors in the external environment that accelerate or delay aging. If so, they might work through the mechanism of changing the rate of production of free radicals or the effectiveness of antioxidant defenses, in which case the environmental explanation would be consistent with and even support the physiological explanation involving free radicals. This reinforcement of one kind of explanation by another can occur when the causal explanations refer to different levels in the system, in this case, the external environment of an organism and its internal physiological state. This idea may seem obvious, but it's amazing how often scientific issues have become hopelessly confused because different researchers have treated their pet hypotheses as competing rather than complementary explanations of a phenomenon. As discussed in Chapter 5, devising alternative hypotheses for a phenomenon and pitting them against each other in rigorous experimental tests is a powerful tool, which can lead to rapid progress in science. However, this method of strong inference can go astray when the alternative hypotheses aren't truly mutually exclusive, as when they operate at different levels.

One of the most interesting environmental factors associated with aging is diet. Researchers began to study the effects of food restriction on the health and longevity of laboratory rodents in the 1930s, and hundreds of experiments since then have pointed to a common pattern that may at first seem surprising (Austad 1997). The pattern in virtually all of these experiments is that rats and mice with restricted diets live longer and healthier lives than mice with "normal" diets. In these experiments, normal diets mean ad lib feeding; that is, subjects are allowed to eat as much as they want. Restricted diets mean that subjects are given 35 to 70% of the calories typically eaten by animals feeding ad lib, although these diets have ample amounts of protein, vitamins, minerals, and other necessary nutrients. Laboratory rodents fed restricted diets live 25 to 40% longer than those allowed to feed ad lib. They are also less susceptible to cancer and infectious diseases, are more vigorous, and have greater memory abilities in old age. One negative effect (from a rodent's point of view) is almost always found in these experiments: animals on restricted diets have markedly lower fertility than those on ad lib diets. This reduction in fertility, associated with greater longevity in laboratory rats and mice, provides a link between the role of diet in aging and some evolutionary explanations of aging that I will discuss later.

How does food restriction prolong life in laboratory rodents? Food intake has many effects on the physiology of rats and mice. Some of these effects are similar between the two species; others are different. There is no general consensus about which of the physiological effects of food restriction is most important for health and longevity, but one intriguing result is that food restriction seems to reduce the production of reactive oxygen species in mitochondria. In some experiments, food restriction also increases the activity of cellular antioxidants such as SOD. How food restriction affects these cellular processes is not known. One logical possibility would be that animals on limited diets have lower metabolic rates, which would explain the reduced production of toxic byproducts of metabolism in mitochondria. Unfortunately, animals on limited diets *don't* have lower metabolic rates, so the mechanism by which food restriction influences the generation of reactive oxygen species, as well as cellular antioxidants, remains unknown. Nevertheless, the link between diet as an environmental factor in aging and free radicals as a molecular mechanism is fairly well established and illustrates the complementarity of causal explanations at different levels.

One limitation of these studies of laboratory colonies of rats and mice is that they probably aren't very relevant to the lives of wild rodents. Laboratory animals have been bred for rapid growth, high fertility, and docility. They eat more and weigh more than wild rodents of the same species; in fact, the physiology of wild rodents may be more similar to that of food-restricted rodents in the laboratory than to that of typical animals in laboratory colonies. However, wild rodents are shorter lived but have greater fertility than food-restricted rodents in the laboratory. Therefore the relationship among diet, aging, and fertility may differ in wild and laboratory rodents. Ultimately, the relationship among these factors under natural conditions is more interesting than the relationship in the laboratory, but also much more difficult to study. The relevance of research on laboratory rodents for relationships between diet and aging in humans is also questionable, although many have wondered if the rodent results might provide hints about how to slow aging in humans (Austad 1997).

GENETIC AND DEVELOPMENTAL CAUSES OF AGING

The physiological and environmental causes of aging discussed so far are factors that influence aging directly and immediately, so they can be called *proximate* causes of aging. Proximate causes of biological phenomena often have their effects within minutes or days. For example, the proximate environmental cause for male birds to advertise their presence in a territory in spring is increased length of days, and the proximate physiological mechanism for the wonderful singing behavior that results is increased levels of testosterone. In aging, damage by free radicals accumulates as animals grow old, and observable effects may not become apparent for many years. Nonetheless, this physiological mechanism can be considered a proximate cause of aging because it occurs during the gradual process of aging itself.

It's also valuable to consider longer term causes of biological phenomena. In particular, the genes that an individual carries and his or her early development may influence the rate of aging. The hypothesis that genetic factors influence aging probably seems perfectly reasonable to you because of the great emphasis on genetics and human health these days (see Chapter 6). The hypothesis that embryonic development affects aging processes that occur many years later may seem more surprising, and I think you will be doubly surprised when I describe how this might work. Genetic and developmental causes of aging and other biological phenomena are especially interesting because they can account for differences among individuals. For example, people die at many different ages. Some people die early in accidents, but even among those who die of natural or intrinsic causes such as cardiovascular disease or cancer, there is wide variation in ages at death. Is this simply chance variation, or are there ways to predict the longevity of individuals?

Researchers have used twin data to study the genetic basis of longevity in humans, using the same methods as described in Chapter 6 for studying the roles of genetic and environmental factors in cancer. In the most detailed study to date, researchers in Denmark compared the life spans of monozygotic twins, who are genetically identical, to those of dizygotic twins, who share half their genes (McGue et al. 1993). They attributed about 33% of the variation in longevity to genetic differences between individuals, comparable to the apparent role of genetic factors in cancer. The average difference in ages at death was 14 years for monozygotic twins versus 19 years for random pairs of individuals from the Danish population. In light of all the pitfalls in interpreting twin studies discussed in Chapter 6, this isn't very strong evidence for an important influence of genetic factors on human longevity.

Another study of unusually old individuals implicated genetic factors more strongly. Thomas Perls and his colleagues (1998) at Harvard Medical School have been intrigued by the lives of centenarians—people who live to the age of 100. As the human population of the world has grown, as health conditions have improved, and as more reliable records of births and deaths have become available, the number of documented centenarians worldwide has increased. The researchers focused on 102 centenarians living near Boston. They compared the ages at death of siblings of these centenarians to ages at death of siblings of a control group of individuals, born at about the same time as the centenarians but who died at the much younger age of 73. Siblings of the centenarians lived longer on average than siblings of those who died at the more typical age of 73. After age 65, siblings of centenarians were about 40% less likely to die each year than siblings of individuals who died at 73, so the chance of living to age 90 was four times as great for siblings of centenarians. Since siblings share half their genes on average, these results suggest that there may be some genetic differences between people who live to very old ages and those who have more normal life spans. It is also worth noting that centenarians are often as healthy as people 10 to 20 years younger. Some of the families studied by Perls's group were amazing. One was a 108-year-old man, his 103- and 97-year-old sisters, four siblings who died after they reached 100, seven first cousins who lived past 100, and 14 additional first cousins who lived into their 90s (M. Duenwald, *New York Times*, 25 December 2001, Section F, p. 1). Perls and his colleagues subsequently analyzed the DNA of a group of people greater than 98 years old and found that they tended to share common genetic information at a specific location on a particular chromosome (Puca et al. 2001). The function of this gene or genes is unknown, but Perls suggests that they may code for enzymes that limit the activity of free radicals, just as in the roundworms discussed earlier.

What about events during embryonic development that may influence aging? One important process that affects female mammals is the production of eggs in the ovaries. Each egg is contained in a sphere of cells called a follicle; all of the follicles that a female mammal will ever have are present at birth. Once she reaches reproductive maturity, groups of follicles go through a series of cyclic changes that culminate in the release of eggs at regular intervals. These cycles are called menstrual cycles in humans and other pri-

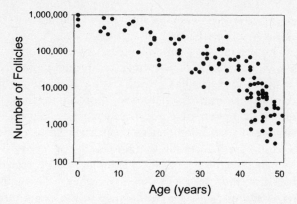

Figure 7.3. Decline in the number of ovarian follicles with increasing age in human females. Estimates of the numbers of follicles were based on autopsies and studies of ovaries removed from women for medical reasons. Note that the vertical axis is logarithmic. As we move up the axis, each labeled tick mark represents 10 times as many follicles as the previous one. On a semilogarithmic graph such as this, a straight line would indicate a constant rate of change in the value of a variable over time. Here, the average number of follicles declines at a rate of about 10% per year from birth to age 38 and then at about 25% per year from age 38 to age 50. Modified from Figure 1 in "Accelerated Disappearance of Ovarian Follicles in Mid-life: Implications for Forecasting Menopause," by M. J. Faddy et al., *Human Reproduction* 7:1342–1346, copyright ©1992 by the European Society of Human Reproduction and Embryology and used with permission of Oxford University Press/Human Reproduction.

mates and estrous cycles in other mammals. In each cycle many follicles and the eggs they contain atrophy and die; one or a small number of follicles develop fully and release eggs that may be fertilized if the female is inseminated by a fertile male at the appropriate time. This process of monthly follicle development is linked to menopause, a major marker of aging in the lives of human females. The age of menopause is quite variable among women, from as young as 35 to as old as 58. This variation is determined by the amount of estrogen produced by accessory cells in the follicles; when the number of follicles remaining in the ovaries declines to about 1,100, estrogen production becomes too low to maintain regular menstrual cycles and menopause begins.

There is wide variation in the number of follicles present in human females at birth, from about 300,000 to 1.1 million. Since the number of follicles declines at a constant or increasing rate with age (Figure 7.3), the age at which a female reaches the threshold for menopause of about 1,100 follicles is directly related to the number of follicles she had at birth. Therefore, the age of menopause is largely determined by the number of follicles present at birth; that is, events during embryonic development influence a fundamental aspect of aging in females 40 to 55 years later. But what are these events? This is where the story, based mostly on studies of mice, gets really amazing (Finch and Kirkwood 2000). It turns out that primordial germ cells (which eventually become eggs) originate outside the embryo itself early in development,[8] then migrate into the embryo and follow a complex path to the site where the ovary is forming (this path takes them through the wall of the gut, for example). The primordial germ cells can be tracked because they have unique granular inclusions. In mice, there are initially about 100 of these cells, which divide to form clusters of cells as they migrate through the embryo. However, the cells in the initial pool do not divide a fixed number of times. When individual primordial germ cells are extracted from developing mouse embryos and grown under identical conditions in culture dishes, some only divide once, to form a colony of two cells; others may divide five or more times, to form colonies containing 32 or more cells. Since conditions in the culture dishes are uniform, the number of times a lineage of primordial germ cells divides is apparently due to chance. In the living mouse embryo, the upshot of this variable proliferation of primordial germ cells is that individual females may have as few as 18,000 or as many as 38,000 of these cells in their ovaries by day 14 of embryonic development. Once in the ovary, each primordial germ cell becomes surrounded by accessory cells to form an ovarian follicle. Many of these follicles die before birth, and this process of death continues as described above until the number of remaining follicles is too small to support continued estrous or menstrual cycles.

To summarize, variation in the reproductive life span in female mammals is strongly influenced by the number of eggs present in the ovaries at birth, which in turn is influenced by random variation in the proliferation of primordial germ cells as they migrate into the ovaries during embryonic development. The evidence for this story comes from observational studies of miscarriages and stillbirths in humans, more detailed experimental studies of mice, and limited research on other mammals (Finch and Kirkwood 2000). In humans, menopause varies by as much as 20 years among women, partly because of random differences in their embryonic development. Finally, there is a hint from a study of centenarians in Massachusetts that women with later menopause may live longer. Perls and his colleagues (1997) found that 20% of the 100-year-old women in their sample had borne children after age 40 (one of these gave birth at age 53), whereas only 5% of women who died at age 73 had given birth after age 40. For this comparison, Perls's group excluded women who never married or who had had hysterectomies before age 35, so opportunities for late childbirth were similar in the two groups. Of course, late births are only indirect evidence of late menopause, but these results at least suggest a possible link between delayed reproductive senescence and longevity in humans.

EVOLUTIONARY CAUSES OF AGING

I hope these examples have convinced you that a comprehensive understanding of aging depends on more than just a physiological or mechanistic explanation. Environmental, genetic, and developmental causes are also important. But all of these kinds of explanations are incomplete because they don't answer one fundamental question: why do different species differ so dramatically in average and maximum life spans? This type of question is the province of evolutionary biologists, who have devoted quite a bit of attention to the problem of aging and have made some contributions that can be integrated very nicely into the other kinds of explanations that we've already discussed.

Evolutionary biologists actually think about two main questions: why does aging exist at all, and why do species differ in their rates of aging? These questions are related because the logic involved in answering the first leads to an approach to answering the second; that is, hypotheses that can be tested by comparing aging in different species.

Aging has been paradoxical for evolutionary biologists because they spend much of their time trying to figure out how traits are advantageous to individuals in an evolutionary sense, so "progressive deterioration of virtually every bodily function over time" (Austad 1997:6) seems inexplicable. How could this possibly be advantageous when evolutionary benefit to individuals means increasing their ability to pass on their genes to future generations? When we think about adaptations that have been molded by natural selection, examples that come to mind include such traits as the great speed and agility of pronghorn antelope, which probably evolved in an environment with several very effective types of predators. If genes influence speed and agility to some extent, if individuals differ in the genes that they carry, and if speedier pronghorn survive longer and therefore have more offspring than slower pronghorn, the frequencies of genes that contribute to speed would increase in subsequent generations of pronghorn populations. This is a straightforward example of how natural selection (differential survival and reproduction of genetically different individuals) can produce evolutionary change. A more mundane example of practical importance to humans is the evolution of resistance to antibiotics in various kinds of microorganisms. In an environment in which antibiotics are used too freely, there is strong selection pressure favoring genes that confer resistance in disease-causing bacteria, which is ultimately detrimental to human health. There is abundant evidence that natural selection has been a pervasive force in the history of life on Earth. How can natural selection explain aging, which is widespread in animals and plants but seems contrary to maximizing reproductive success by surviving and remaining healthy in order to reproduce as long as possible?

The answer to this puzzle was first articulated by the British biologist Peter Medawar in 1952. The approach to aging taken by Medawar and many subsequent evolutionary biologists was fundamentally different from that taken by various researchers to most of the problems discussed so far in this book. To this point, I've emphasized using concrete evidence to solve problems in biology. Often the best evidence has come from rigorous experiments. By contrast, most of the initial work by Medawar and others was purely theoretical. These researchers essentially did *thought experiments* to come up with a logically coherent explanation of the paradox of aging. If this seems dissatisfying compared to real experiments, recall that much of Einstein's revolutionary contributions to physics used exactly the same approach. In Einstein's case, people eventually figured out observations to make and experiments to do to test his novel ideas; the same has occurred in aging research.

The key to an evolutionary understanding of aging is the distinction between extrinsic and intrinsic sources of mortality. Suppose there is a baseline rate of extrinsic mortality due to predation, disease, and accidents that is independent of age. Superimposed on this baseline rate is additional mortality due to factors intrinsic to individuals in a population. In humans and some other animals, these factors include things like cancer and cardiovascular disease. In grazing mammals such as deer, they include wear on the grinding surfaces of the cheek teeth from years of chewing abrasive vegetation. For other species, different intrinsic factors may contribute to mortality. In almost all cases, however, the rate of intrinsic mortality increases with age, unlike that of extrinsic mortality.

Medawar's (1952) evolutionary insight about aging came from thinking about the consequences of a constant rate of extrinsic mortality. Imagine a cohort of 1,000 mice born at the same time. If the extrinsic mortality rate is 10% per month (remember that this is a thought experiment, so we can use any figure we want), about 900 mice would survive to be 1 month old, 810 would survive to be 2 months old, and so on. About 280 mice would survive for a full year, about 80 for 2 years, and fewer than five for 3 years[9] (the maximum life span of house mice in captivity is 4 years, so our assumption of an extrinsic mortality rate of 10% per month in nature may not be too unrealistic). Since the probability that a mouse would survive to be 3 years of age is very low, *even in the absence of any intrinsic mortality associated with aging*, natural selection against any mutations that cause deterioration at old ages will be weak. Mice that carry such mutations might have a higher probability of dying at or beyond age 3 than those that don't, say 15% per month versus 10% per month for those that experience only extrinsic mortality. However, since the chance of surviving to age 3 in the first place is less than 0.5%, most of the reproduction by mice that carry the deleterious gene and mice that don't will have already taken place. Therefore, any potential disadvantage of the deleterious gene in the face of natural selection will be very slight, these kinds of genes won't be readily eliminated from populations, and we have an evolutionary mechanism for aging: the hypothesis that aging is due to the accumulation of genes that have deleterious effects at older ages.

A human example may make this idea less abstract. Huntington's disease kills people in middle age after 10 to 20 years of progressive muscular and mental deterioration. One of the best known victims was the famous American folk singer Woody Guthrie, who died at age 55 after spending the last 13 years of his life in a hospital. A single dominant gene causes Huntington's disease; that is, a person with one copy of the gene and one copy of its normal counterpart is afflicted. Such a person would pass the gene for Huntington's disease to half of his or her offspring on average. In fact, because Huntington's disease strikes in middle age, there's a good chance that an affected person would not come down with the disease until after having children. For example, Woody Guthrie's mother had five children before being stricken, and he had six. Despite its devastating effects, Huntington's disease is relatively more common than other fatal and even nonfatal genetic diseases, occurring at a frequency of one in 15,000 among people of European descent (Austad 1997). By contrast, inherited retinoblastoma, which I discussed in Chapter 6, occurs in only about one in 25,000 people, even though it is nonfatal. A major difference between these two diseases is that retinoblastoma affects people in infancy, before they have had a chance to reproduce, whereas Huntington's disease affects people after most or all of their opportunities for reproduction have passed. The blindness or other visual problems caused by retinoblastoma before the age of modern medicine would probably have affected the reproductive success of carriers of this gene, even without killing them, whereas Huntington's disease invariably causes premature mortality but has little impact on reproductive success because it occurs in middle age. Therefore, natural selection against Huntington's disease would not be as strong as that against retinoblastoma, even though the former is always fatal and the latter is not. As a result, the frequency of retinoblastoma is substantially lower than the frequency of Huntington's disease.

The mutation-accumulation hypothesis illustrated by the examples of hypothetical life spans in mice and Huntington's disease in humans is one of three major evolutionary hypotheses for aging. The second, closely related to the first, is that mutations that have beneficial effects early in life but detrimental effects later will be favored by natural selection. This hypothesis has a wonderful name, antagonistic pleiotropy. Genes that have multiple effects are pleiotropic; if these effects are positive at one stage of life and negative at another, they can be said to be antagonistic. A plant or animal's lifetime reproductive success is likely to be increased by a gene that helps it survive better or reproduce more in the early part of its life. If the same gene adversely affects its survival or reproduction later in life, its lifetime reproductive success will be diminished, assuming it lives that long. However, the net effect of the gene is likely to be positive because there is only a low probability that the individual will survive to old age anyway, so the effect of the gene late in life is relatively unimportant compared to its effect early in life.

The third specific hypothesis for the evolution of aging is the disposable-soma hypothesis. "Soma" refers to the nonreproductive tissues of the body. These are disposable in the sense that their evolutionary function is to maintain an organism in good condition until it has an opportunity to reproduce. As the cumulative probability of surviving declines because of extrinsic mortality (recall the mouse example discussed earlier), the somatic tissues become more and more disposable because the chance of being alive, and therefore being able to reproduce at older ages, is low. Since an organism would need to invest metabolic resources to maintain its soma for a long time, it may be advantageous to invest these resources in early survival and reproduction instead. For example, long life might depend on devoting metabolic resources to mechanisms for repairing damage to DNA, proteins, and lipids by free radicals. But if extrinsic mortality is relatively high, it might be more beneficial to devote these resources to increasing reproductive success in early years. By "more beneficial," I mean that this strategy might contribute to greater net reproductive success and therefore be favored by natural selection, so individuals that followed the early reproduction strategy would leave more descendants than individuals that followed the soma-maintenance strategy.

To recapitulate, theoretically inclined evolutionary biologists have proposed three hypotheses for the evolution of aging: mutation accumulation, antagonistic pleiotropy, and disposable soma. These hypotheses are difficult to distinguish because they are not mutually exclusive and they make similar predictions. It will probably require detailed genetic studies to discriminate among them in particular cases. However, extrinsic mortality plays a key role in all three hypotheses, and the fundamental prediction of all three is that the rate of aging should be higher in species with higher levels of extrinsic mortality than in species with lower extrinsic mortality. Where extrinsic mortality is high, aging should be rapid because natural selection favors early reproduction at the expense of traits that might delay aging. Where extrinsic mortality is low, reproduction at older ages is more likely, so natural selection should favor traits that protect against aging and promote survival.

Steven Austad (1993) reported one of the first tests of a correlation between extrinsic mortality and rate of aging under natural conditions. He compared two populations of Virginia opossums, one on Sapelo Island, off the coast of Georgia, the other at the Savannah River Ecology site in South Carolina. These aren't different species, but Austad had good reason to believe that they were genetically isolated from each other. Opossums are midsized mammals, but because they don't have particularly good defenses against predators they tend to reproduce early and often and to be short lived. In an earlier study in Venezuela, Austad found that opossums showed signs of senescence, such as cataracts, arthritis, and hair loss, at about 18 months and never lived beyond age 2. When he started working in the southeastern United States, he found that opossum predators were completely absent from Sapelo Island. Since predation is a major source of extrinsic mortality for opossums, this led Austad to predict that aging should be slower on Sapelo Island than on the mainland.

Austad used some standard methods of wildlife biology and some innovative approaches to test this prediction. He collected data on survival and reproduction by capturing females and fitting them with radio collars. Each radio collar emits a unique frequency, so Austad could use a receiver and portable antenna to relocate these opossums. The radio collars also had mortality sensors, which changed the frequency of the emitted signal when an opossum died. Therefore, Austad could determine the ages at death of his study animals fairly precisely. In addition to this standard technique of collecting demographic data on wild animals, Austad used a relatively novel method to assess one physiological aspect of aging. He collected a few tendon fibers from the tail each time an opossum was caught. These contain collagen, which is the most common protein in mammals and other vertebrates. As animals age, the chains of amino acids making up collagen become cross-

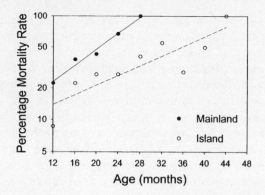

Figure 7.4. Annual probability of death of female opossums studied by Steven Austad on Sapelo Island and at a mainland site near Savannah, Georgia. There were no predators on the island but abundant predators on the mainland. The symbols show estimates of the mortality rate at 4-month intervals based on radiotelemetry. The lines are best-fitting regression lines for the two populations. These are determined statistically and show overall trends across all ages for each population. The slopes of these lines can be expressed as mortality-rate-doubling times as described in the text. Modified with permission from Figure 1 in "Retarded Senescence in an Insular Population of Virginia Opossums (*Didelphis virginiana*)," by S. N. Austad, *Journal of Zoology, London* 229:695–708, copyright ©1993 by Cambridge University Press.

linked (this is one manifestation of free-radical damage), which makes collagen fibers less flexible. This loss of flexibility causes fibers from tendons in the tails of older opossums to last longer before breaking when a weight is attached to them under standard conditions in the laboratory.

Austad found that the average life span of female opossums was 24.6 months on Sapelo Island and 20 months on the mainland. There was an even bigger difference in maximum life span: 45 months for the island population and 31 months for the mainland population. These differences were due not only to the lack of predators on Sapelo Island but also to differences in the rate of aging of the two populations. Specifically, the mortality rate at 12 months, when females reach reproductive maturity, was about 22% per month for mainland opossums and 14% per month for island opossums. This illustrates the greater extrinsic mortality on the mainland. Beyond 12 months, the mortality rate doubled about every 7.6 months on the mainland and 12.8 months on Sapelo Island. This illustrates the slower rate of aging of island opossums (Figure 7.4). In addition, the breaking time for tail tendon fibers increased more slowly with age for the island population than for the mainland population. Finally, island females had smaller litters (5.7 young on average) than mainland females (7.6 young). Since island females were likely to live longer than mainland females, they had a better chance of having more litters before they died, so having large litters early in life was less important for their lifetime reproductive success.

Comparative studies like this can be challenged because many other factors besides the one thought to be responsible for a pattern may differ between sites. In this case, Sapelo Island might have a different climate, different food resources, or different pathogens or parasites than the mainland, and one of these factors, rather than the absence of predators, might account for the differences in aging and reproduction between island and mainland opossums. If it were possible to design an experiment in which several islands, all lacking predators, were selected and then predators were introduced to half of these islands, we could control for the effects of these other factors by randomly picking the islands to get the predators. But this experiment probably wouldn't be very definitive even if it were practical and could be justified ethically because it might take many generations for a difference in the rate of aging to evolve in the opossums in response to a change in extrinsic mortality. If we didn't see a difference in a reasonable time period for the study, an advocate for the evolutionary hypothesis could argue that we just hadn't waited long enough for the predicted evolutionary change to occur. So the comparative approach, despite its flaws, is about the best we can do if we want to tackle an evolutionary question by using a species with a moderate to long generation time under natural conditions.

In fact, Austad considered these alternative explanations for demographic differences between island and mainland opossums and was able to reject them. Climate was similar at the two sites, although average temperature was slightly higher on the island. In experiments with mice, tail tendon fibers broke more quickly at higher temperatures. If temperature explained the difference in breaking time between island and mainland opossums, the time should have been faster for the island animals, just the opposite of the actual results. There were slightly higher frequencies of ticks on the ears of island opossums, although mortality rates were lower for this population. Population density of opossums was higher on the island, suggesting the possibility of food shortage. In lab experiments with mice and rats, as described above, diet restriction can delay senescence, and this mechanism might explain the slower rate of aging of island opossums. But there were no differences in body mass index (a standard measure of fat content of an animal) or blood glucose levels between island and mainland opossums, although these two measurements typically differ between mice or rats on restricted and normal diets. Therefore, Austad's study provides fairly persuasive evidence for the relationship between extrinsic mortality and rate of aging predicted by the evolutionary theory of senescence.

Robert Ricklefs (1998) of the University of Missouri did a broader comparison of multiple species of birds and mammals to test the prediction that aging occurs more quickly in species with higher extrinsic mortality. This kind of analysis requires data on mortality rates of different age classes in a population. The minimum mortality rate, which typically occurs at about the age of sexual maturity, is assumed to largely represent extrinsic mortality; a graph of increasing mortality rate at ages beyond sexual maturity (similar to Figure 7.1) is used to estimate the rate of aging. For 18 species of birds and 27 species of mammals ranging in size from indigo buntings to swans and rabbits to hippos, Ricklefs found that the rate of aging was higher for species with greater extrinsic mortality. For example, the extrinsic mortality rate for hippos was about 3% per year, the rate of aging was 0.05, and the maximum age was 43 years, whereas for zebras the extrinsic mortality rate was 6% per year, the rate of aging was 0.10, and the maximum age was 23 years.

In this chapter we have seen that causes of aging can be viewed through different lenses. There are physiological mechanisms of aging, including fundamental cellular and molecular processes; environmental causes such as diet; and genetic and developmental differences among individuals that influence aging. Moreover, aging makes sense from an evolutionary perspective, even though it initially seemed paradoxical.

Let me close with two puzzles about aging that are active topics of current research. The first is that many birds are remarkably long lived compared to mammals of similar size. For example, the tiny broad-billed hummingbird can live to 14 years of age in nature, and canaries live for more than 20 years in captivity. Compare these values to house mice, which live at most 4 years in captivity, or even your pet dog, which will probably die before your pet canary of the same age. One surprising thing about the longer lives of birds is that they have higher metabolic rates than mammals, so generation of reactive oxygen species, which damage DNA, proteins, and lipids, should be higher in birds. There is some evidence that birds age slowly because they have specialized antioxidant defenses that counter the effects of these reactive oxygen species. In fact, Donna Holmes and Steven Austad (1995) suggest that birds and other long-lived species may be model organisms for learning more about physiological mechanisms of aging. In other words, the comparative approach, rooted in questions about the evolution of aging, can contribute to increased understanding of physiological, environmental, genetic, and developmental aspects. However, the evolutionary reason that birds tend to age more slowly than mammals of similar size is not completely clear. Various researchers have suggested that flight in birds makes them less vulnerable to predation, which means they should have lower extrinsic mortality, which should be associated with lower rates of aging. The fact that bats live longer than terrestrial mammals of similar size is consistent with this hypothesis, but more focused tests of the hypothesis are needed. These might include the determination of whether predation rates really are lower for flying birds and mammals than for nonflying ones, as well as comparisons of aging in different species of birds or bats that experience different levels of extrinsic mortality.

A second fascinating puzzle about aging is why human females live for many years after menopause. This is very unusual in animals. One possible reason is that postmenopausal females continue to contribute to their genetic legacy by helping to raise their grandchildren, so living beyond menopause could be favored by natural selection. Alternatively, life after menopause might be an artifact of medical advances in modern society and have no evolutionary significance (Sherman 1998; Shanley and Kirkwood 2001). Anthropologists have much to contribute to solving this puzzle, but I predict that detailed population and behavioral studies of other species, such as whales and elephants, in which females live beyond menopause may be especially valuable.

RECOMMENDED READING

Austad, S. N. 1997. *Why we age: What science is discovering about the body's journey through life.* Wiley, New York. An engagingly written introduction to many different perspectives on aging.

Holekamp, K. E., and P. W. Sherman. 1989. Why male ground squirrels disperse. *American Scientist* 77:232–239. An illustration of the different levels of analyzing causation, which is the main theme of this chapter.

Kirkwood, T. B. L., and S. N. Austad. 2000. Why do we age? *Nature* 408: 233–238. An introduction to a recent series of articles about different aspects of aging.

Ricklefs, R. E., and C. E. Finch. 1995. *Aging: A natural history.* Freeman, New York. Like Austad (1997), an excellent overview of research on aging.

Chapter 8

How Does Coffee Affect Health?
Combining Results of Multiple Studies

We are bombarded with information and advice about health and nutrition. Here is a sample of recent headlines: "Something Special in That Glass of Wine" (*New York Times*, 26 September 2000); "Warding off Parkinson's with Caffeine" (*New York Times*, 24 October 2000); "Conflicting Views on Caffeine in Pregnancy" (*New York Times*, 17 July 2001); "Coffee in the Afternoon a Memory Boost for Elders" (*Health & Medicine Week*, 7 January 2002); "A Dose of Red Pepper to Sooth Gastric Distress"(*New York Times*, 2 April 2002). Stories like these usually report the conclusions of some new scientific study, but these conclusions often contradict previous research, which is partly what makes them newsworthy. How should we judge these kinds of stories, which may encourage us to fine-tune our eating habits or lifestyle? More generally, how does consensus develop among scientists and physicians so that conjectures and hypotheses become generally accepted principles? How often do such principles get overturned as a result of further research, and how does this process of *paradigm shift* occur?

We have discussed several different kinds of evidence used to test hypotheses in biology and medicine, ranging from correlational studies and other observational research to various kinds of experiments. The credibility of an hypothesis is often enhanced if different types of evidence are consistent with it. For example, Pieter Johnson and his colleagues (1999) tested the parasite hypothesis for limb deformities in frogs by a lab experiment and complementary field observations and got congruent results from these approaches (Chapter 4). Similarly, the credibility of the physiological hypothesis that aging is due to molecular byproducts of cell metabolism benefits from the fact that this mechanistic hypothesis is compatible with environmental, genetic, developmental, and evolutionary explanations of aging (Chapter 7).

Another important element in the attainment of consensus in science is replication of experiments or observational studies to test particular hypotheses. The story of cold fusion is a classic example of this. In 1989, a pair of researchers at the University of Utah reported a surprising result—fusion of deuterium atoms to form helium and release energy at room temperature. Since deuterium can be extracted from seawater, this process promised unlimited and inexpensive energy if it worked. Because of the potential importance of cold fusion to society and because it contradicted conventional wisdom that nuclear fusion requires extremely high temperatures, such as those in the interior of the sun, chemists and physicists rushed to replicate the work of the Utah researchers in their own labs. No one was able to do so, and cold fusion was rapidly discredited (Park 2000).

Replication is also important for more mundane questions; for example, does vitamin C protect against colds? In cases like this, however, results often aren't as clear as in the unsuccessful attempt to replicate cold fusion. Various researchers may do multiple experiments, producing some results that are consistent with an hypothesis, other results that are inconsistent, and still others that are inconclusive. Two characteristics of medical research make this especially common. There is great interest in preventing and treating disease, so funding is available for many studies to test treatments or prevention strategies. However, tremendous variation among the human subjects of medical research makes it difficult to get consistent results in multiple studies of the same question.

This uncertainty about the results of research creates a confusing situation for consumers of medical research who wish to apply these results to improving health. These consumers include physicians, public-health workers, and members of the general public who yearn to live longer and healthier lives. Despite the uncertainty of many research results in nutrition and medicine, doctors often subscribe to a *standard of care* for their healthy patients, which includes generally accepted recommendations about diet, exercise, regular screening tests for certain diseases, and preventive medications. For example, physicians routinely advise patients to limit their intake of caffeinated coffee, to have annual mammograms or tests for prostate cancer after a certain age, and to take aspirin regularly if they have certain risk factors for cardiovascular disease. What is the source of these standards of care when there are conflicting studies of the physiological effects of caffeine, of the costs and benefits of mammography and other screening tests (Grimes and Schulz 2002), and of the efficacy of various medications? Do physicians reach a consensus about standards of care based on a few exemplary studies that are more persuasive because they seem more rigorous than other studies of the same phenomena with different results (LeLorier et al. 1997)? How about combining results of different studies in some way to reach a general conclusion? How important for agreement about standards of care are sociological, economic, and political factors such as promotion of drugs by pharmaceutical companies?

The potential effects of coffee on health provide a good illustration of these issues. According to Robert Superko and his colleagues, "Fifty-six percent of the adult US population consumes an average of 3.4 cups . . . of coffee per day, and as an import, coffee is second only in importance to oil" (1991:599). Coffee is used even more widely in Europe, where decaffeinated coffee is much less popular than in the United States. For example, 94% of adults in the Netherlands drink coffee daily; only 4% of this is decaffeinated compared to 20% in the United States (van Dusseldorp et al. 1989). Epidemiologists and medical researchers have investigated many proposed health effects of coffee, ranging from increasing the risk of bladder cancer to protecting against colorectal cancer to causing higher blood pressure and greater levels of cholesterol in the blood, which might contribute to cardiovascular disease. I searched an online database of biomedical literature maintained by the National Library of Medicine and found almost 3,000 articles published between 1990 and 2000 with coffee or caffeine in their titles.

CAFFEINE AND BLOOD PRESSURE: AN EXEMPLARY STUDY

Before tackling the integration of results from such an abundance of research, I need to introduce some fundamental ideas of statistics that relate to testing hypotheses in individual experiments. I will use a study by four Dutch researchers to illustrate these ideas. Marijke van Dusseldorp and her colleagues (1989) compared the effects of regular and decaffeinated coffee on blood pressure in an experiment with 45 subjects from the town of Nijmegen. They advertised for volunteers in local newspapers and at the University of Nijmegan. From 150 people who were willing to take part in the experiment, the researchers selected 45 who were between 17 and 45 years old, did not smoke or work at night, were not pregnant, were generally healthy, and drank coffee regularly. They used a standard type of experimental design called a *crossover trial*. Twenty-three of the subjects drank regular filtered coffee for 6 weeks, then decaffeinated filtered coffee for 6 weeks. The remaining 22 subjects drank decaffeinated coffee for the first 6 weeks and regular coffee for the next 6 weeks. Subjects were randomly assigned to these two groups in such a way that each group had similar numbers of males and females, a similar range of ages, and similar baseline blood pressure measurements.

Each subject was given a supply of coffee in individual packages, a coffee maker, and instructions on how to prepare the coffee. They were asked to drink 2 cups of coffee before noon, 1 cup in the afternoon, and 2 cups in the evening. They were also asked not to drink tea or use other substances containing caffeine, although limited use of chocolate was permitted. The researchers collected two blood samples during the trials to check for caffeine and thus to be sure all subjects were following instructions. The subjects were also given a device for measuring their own blood pressure, and they did so five times on 1 day each week.

This was a randomized, double-blind experiment. I have described the randomization process above. Two things made it a double-blind experiment. First, the researchers did not know which subjects were in each group (regular coffee for 6 weeks and then decaffeinated, or decaffeinated coffee first and then regular) because assistants who were not involved in processing or analyzing any of the data did the random allocation of subjects to the two groups and distributed the packages of coffee. Second, at least according to the researchers, the subjects were unable to guess when they were drinking regular coffee and when they were drinking decaffeinated (evidently the caffeine-deprived subjects didn't suffer the headaches and lethargy often experienced by caffeine addicts who don't get their daily fix).

Figure 8.1, from the article by van Dusseldorp and her colleagues, nicely summarizes one of the key results of their work. Editors rarely allow authors to include data for individual subjects in their articles, instead requiring the results to be summarized as average values in tables or figures (sometimes the large number of subjects in a study precludes publication of individual data; sometimes the economics of publishing stands in the way). But the editor of the journal *Hypertension* allowed the Dutch researchers to publish this graph, which shows the difference in systolic blood pressure for each subject on the two types of coffee.[1] These measurements were taken during the last 2 weeks of the 6-week period on each type of coffee, so they represent a month of acclimation to each type. The leftmost bar shows that average systolic blood pressure for one subject was almost 7 units less when drinking regular coffee than when drinking decaffeinated. The rightmost bar shows a difference of about 9 units in the opposite direction; that is, the average systolic blood pressure for this subject was 9 units greater when drinking regular coffee than when drinking decaffeinated. Take some time to examine this figure and think about the amount of individual variation that exists among the 45 subjects of this experiment.

How can we best summarize these results? One possibility is to notice that a majority of individuals had positive values, which means higher systolic pressure on regular coffee than on decaffeinated. But 15 of the 45 subjects had negative values, corresponding to higher pressure on decaffeinated. A second possibility is to compute the average difference between regular and decaffeinated coffee for the 45 subjects, which is 1.5 units; that is, systolic pressure is an average of 1.5 millimeters of mercury higher on regular coffee than on decaffeinated. This is not a very large difference. Is it large enough to conclude that drinking decaffeinated coffee is likely to lower blood pressure somewhat for the average person?

A common way to answer questions like this is to estimate the probability that a difference of this magnitude could arise purely by chance. If this probability is relatively low, we may be justified in concluding that caffeine has an effect on blood pressure, although we can't absolutely exclude the possibility that the difference of 1.5 units in systolic pressure was due to chance. Of course, this line of reasoning depends on assuming that the results are unbiased. If flaws in the experiment led to biased results, there's little point in thinking

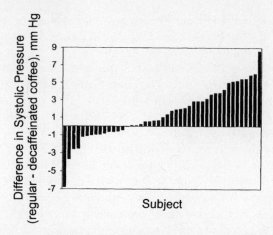

Figure 8.1. Difference in average systolic blood pressure when drinking regular and decaffeinated coffee for 45 subjects studied by van Dusseldorp and her colleagues in the Netherlands. Each subject drank regular coffee for 6 weeks, then decaffeinated coffee for 6 weeks, or vice versa. Blood pressure was measured on 2 days during the last 2 weeks of each 6-week portion of the experiment. Each bar is the average for an individual subject. Positive values mean that systolic pressure was higher with regular coffee; negative values mean that systolic pressure was higher with decaffeinated coffee. Modified with permission from Figure 1 in "Effect of Decaffeinated Coffee versus Regular Coffee on Blood Pressure: A 12-week, Double-blind Trial," by M. van Dusseldorp et al., *Hypertension* 14:563–569, copyright ©1989 by Lippincott Williams & Wilkins.

about how chance might have influenced these results. For example, suppose the researchers had instructed all subjects to drink decaffeinated coffee for the first 6 weeks and regular coffee for the next 6 weeks and there was a series of heavy snowstorms during the period when everyone was drinking regular coffee. The bad weather might have caused increased stress and higher blood pressure, leading to an inaccurate, or biased, estimate of the effect of caffeine on blood pressure.

There are several conceivable ways in which chance might affect the results of an experiment like this. A person's blood pressure is influenced by many factors—such as time of day, temperature, diet, exercise, and stress—and may vary quite a bit from hour to hour or even minute to minute. It is also difficult to measure blood pressure accurately and consistently; even if a person's blood pressure remained constant for a few minutes, successive measurements might differ because of imprecision of the measuring process. Any of these sources of variation might cause a subject in the experiment to have different measurements of blood pressure during the last two weeks of the decaffeinated phase of the experiment than during the last two weeks of the regular coffee phase, even if caffeine has no effect on blood pressure. Moreover, individuals differ in their responses to caffeine and other factors that affect blood pressure. If the experiment were repeated with another group of

45 volunteers, these individual differences would probably cause the average effect of the type of coffee to differ from the 1.5 units that was found for the 45 subjects that actually participated. In other words, the estimated effect of caffeine on blood pressure depends on chance at two levels: variation within individuals in measurements of blood pressure due to factors besides caffeine, and differences between individuals who participated in the experiment and others who might have been selected from the population of healthy young adults in the Netherlands.

We can use a thought experiment with the data in Figure 8.1 to estimate the effect of the first kind of chance variation. Imagine that caffeine has no effect on blood pressure (the hypothesis of no effect is called the *null hypothesis*) and that the lengths of the bars in Figure 8.1 simply represent differences due to all of the other factors that affect pressure. If this were the case, each bar would have an equal chance of being above the horizontal axis, representing no difference in pressure, or below this horizontal axis. For example, the left-most subject in Figure 8.1 had a systolic pressure while drinking regular coffee that was almost 7 units less than he or she had while drinking decaffeinated coffee. If this wasn't due to the difference in the type of coffee, but rather to other factors that cause day-to-day variation in this subject's blood pressure, the difference would be just as likely to be in the opposite direction, seven units greater on regular than on decaffeinated coffee.

Suppose we randomly select a direction from the horizontal axis for the pressure difference for each of the 45 subjects and then calculate the average difference. Figure 8.2 shows one example of a pattern we might get. In this case, 23 of the subjects have a positive difference in pressure (regular greater than decaffeinated), 22 have a negative difference (decaffeinated greater than regular), and the average difference is 0.3 units. Compare these numbers to the actual results: 30 subjects with a positive difference, 15 with a negative difference, and an average of 1.5 units.

I repeated this process 10,000 times and got only 15 cases in which the average difference in pressure was greater than 1.5 units in either direction. This implies that, if caffeine has no effect on blood pressure, the chance of getting a difference as great as 1.5 units between drinking decaffeinated coffee and drinking regular coffee is only 15/10,000, or about 0.2%. In other words, the results of the Dutch researchers shown in Figure 8.1 are not likely to be due to chance variation in blood pressure within individuals, and we have some justification for rejecting the null hypothesis that caffeine doesn't affect blood pressure.

How about the other major source of chance variation, differences between individuals who actually participated in the experiment and other potential subjects? The most common way to measure the effect of this kind of random variation is to assume that the subjects in the experiment were a representative sample of a particular population and that a graph of blood pressure differences for members of this population would have a particular shape called a *normal distribution*. A normal distribution is a symmetrical, bell-shaped curve showing the number of expected cases in a population for each

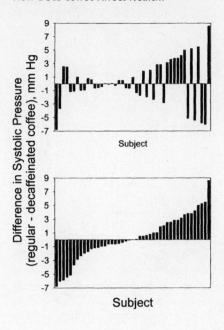

Figure 8.2. One trial in which the direction of the bar for each subject illustrated in Figure 8.1 (above or below the horizontal axis) was randomly determined. Panel A shows the random results that were obtained; panel B shows the same results arranged in order from the smallest (most negative) to the largest value.

value of a variable of interest. The peak of the curve occurs at the average value of the variable; the breadth of the curve depends on the amount of variation in values of that variable in the population. Figure 8.3 is a histogram of the actual differences in systolic pressure while drinking the two types of coffee for the 45 subjects in the Dutch experiment. The average or mean difference was 1.5 units; the standard deviation, a typical measure of variability in the data, was 3.0. Figure 8.3 also shows a normal curve of difference in pressure based on this mean and standard deviation. Since the shape of the histogram approximates the shape of the normal curve in Figure 8.3, blood pressure differences for the experimental subjects were fairly close to being normally distributed. Furthermore, if the *sample* of subjects used in the experiment was representative of the population as a whole, the height of the normal curve for each difference in pressure represents the proportion of individuals in the *population* who would be expected to have that value.

I've been discussing this hypothetical population in fairly abstract terms so far. Who actually belongs to this population? This depends on how broadly the Dutch researchers intended to generalize the results of their experiment. The subjects included both males and females, so clearly the population of interest to the researchers included both sexes. All subjects were healthy adults between 17 and 45 years of age, so we would not want to generalize these results to senior citizens or to people with hypertension. Although all of the subjects lived or worked in the town of Nijmegen, the researchers presumably assumed that these subjects were representative of all healthy young adults in the Netherlands, of healthy young adults in northern Europe, or perhaps of healthy young adults in industrialized countries worldwide.

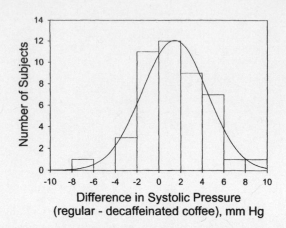

Figure 8.3. Summary of the results shown in Figure 8.1 for the 45 subjects in the experiment by van Dusseldorp and her colleagues (1989). The height of each bar of the histogram represents the number of subjects whose difference in systolic pressure with regular and decaffeinated coffee was within the range shown on the horizontal axis. For example, systolic pressure was 4 to 6 units greater on regular coffee than on decaffeinated coffee for seven subjects. The curve shows a normal distribution fitted to these data with an average of 1.5 units and a standard deviation of 3 units.

What if a different sample of 45 subjects had been selected from the population represented by the normal curve in Figure 8.3? The average blood pressure difference would probably not be exactly the same as the value of 1.5 for the individuals who actually participated in the experiment, but how much would the average for this hypothetical sample depart from the average actually obtained? If we did this thought experiment many times, we could get an idea of the range of possible values for the average difference in blood pressure for many possible samples. It's relatively easy to do so by using a computer program that simulates drawing 45 values at random from a normal distribution, calculating the average of these 45 values, and repeating this process 10,000 times. I did this and obtained the results shown in Figure 8.4.

In only three of these 10,000 trials was the average difference in pressure less than zero, which would mean greater systolic pressure on decaffeinated coffee than on regular coffee. In 95% of the trials, the average difference was between 0.6 and 2.3 units, as illustrated by the dotted lines in Figure 8.4. This range is called the *95% confidence interval* for the difference in systolic pressure while drinking regular and decaffeinated coffee; that is, we can be 95% sure that the average difference in systolic pressure on the two types of coffee is between 0.6 and 2.3 units for the population represented by these subjects. Because this confidence interval includes only positive values, our analysis suggests that the modest average reduction of 1.5 units in systolic pressure from drinking decaffeinated coffee in the Dutch experiment was not an artifact of chance in the selection of the subjects. Of course, this depends

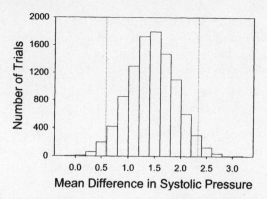

Figure 8.4. Summary of 10,000 trials in which values for 45 hypothetical subjects were randomly selected from the normal distribution illustrated in Figure 8.3. For each trial, the average difference in systolic pressure was calculated; the histogram shows the distribution of these average values. Ninety-five percent of the values fall between the dotted lines, so the 95% confidence interval for the average difference in systolic pressure with regular versus decaffeinated coffee is from 0.6 to 2.3 units.

fundamentally on the assumption that the 45 subjects were representative of the population at large. If they weren't, we couldn't derive the normal curve in Figure 8.3 to represent the population, and so we couldn't make the graph in Figure 8.4 from which the confidence interval was calculated.

The Dutch researchers found a slightly smaller reduction in average diastolic blood pressure on decaffeinated coffee, 1 unit compared to 1.5 units for systolic pressure. The 95% confidence interval for the difference in diastolic pressure was 0.2 to 1.8 units. They also found that it didn't make any difference if individuals drank decaffeinated coffee for 6 weeks, then regular coffee, or vice versa. Therefore, they rejected the null hypothesis that caffeine has no effect on blood pressure and concluded that switching to decaffeinated coffee would probably produce a small drop in blood pressure, at least for healthy young adults. But what is the biological significance of this result? The average baseline blood pressure of the subjects in this experiment was 124/76, so decreases of 1.5 units in systolic pressure and 1 unit in diastolic pressure would reduce normal values by less than 2%. Is this small change likely to decrease the risk of heart attack, stroke, or other cardiovascular disease? Geoffrey Rose (1981) reported that about 70% of the deaths from heart disease and stroke in a sample of British men occurred in those with diastolic pressure below 110 mm Hg. Although an individual's chance of dying was much greater if his blood pressure was greater than 110, a small fraction of the men were in this category, so many deaths due to cardiovascular disease occurred in men with lower blood pressure. This implies that as many heart attacks and strokes might be prevented by decreasing average blood pressure in the population as a whole by two to three units as by treating only individuals with very high blood pressure with expensive medications. This idea has come to be known as the *paradox of prevention:* for any one individual, the decreased probability of dying from cardiovascular disease because of switching from regular to decaffeinated coffee is probably minuscule, but if everyone switched the reduction in the death rate might be substantial (Rose 1992).

META-ANALYSIS: COMBINING RESULTS OF MULTIPLE STUDIES

The Dutch study was well designed and produced clear results, but many other researchers have also done experimental studies of the effects of caffeine on blood pressure, using different experimental designs and subjects from different populations. Some had fewer subjects than the Dutch study; others had larger sample sizes. How consistent were the results of these various studies? Is the effect of caffeine on blood pressure replicable across a range of different conditions, or were there some unique features of the Dutch study that produced an effect specific to those conditions? Sun Ha Jee and four colleagues from Johns Hopkins University in Baltimore, Maryland, and Yonsei University in Seoul, South Korea, reviewed the literature in 1999 in an attempt to answer these questions.

The first experimental study of coffee consumption and blood pressure took place in 1934, and Jee's group found a total of more than 36 studies that had been published by 1999. Many of these experiments, however, had limitations that precluded quantitative analysis. For example, some studies compared a group of people drinking coffee and being treated with medication for hypertension with a control group not drinking coffee but also not being treated with such medication. Clearly, any difference in blood pressure between these two groups could not be attributed to caffeine alone. Because of flaws like this in most of the experiments, Jee and his colleagues were able to include only 11 studies in their analysis. These were quite diverse in experimental design and in characteristics of the subjects. Six of the studies used a crossover design in which each subject spent a period of time drinking regular coffee and another period of time in a control condition, as in the study by van Dusseldorp and her colleagues (1989). The other five studies used separate treatment and control groups. In some cases, the control condition entailed drinking decaffeinated coffee, as in the Dutch study; in other cases, the control condition entailed no consumption of coffee. One study used subjects with hypertension; the remaining studies used subjects with normal blood pressure. The shortest study lasted 2 weeks; the longest, about 11 weeks. The average age of subjects in the 11 studies ranged from 26 to 56, and the sample sizes ranged from 8 to 99.

Figure 8.5 shows the average effect of caffeine on blood pressure for these 11 studies, as well as confidence intervals for each average effect. Caffeine was associated with increased systolic pressure in nine studies and with increased diastolic pressure in nine studies, although these weren't the same nine studies. The 95% confidence intervals did not extend to zero in six cases for both systolic and diastolic pressure. This counts as evidence that caffeine affects blood pressure. But the remaining five studies had 95% confidence intervals

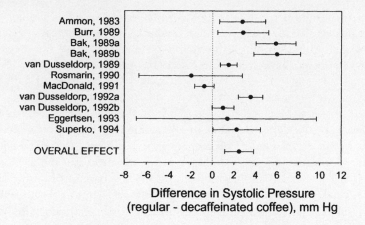

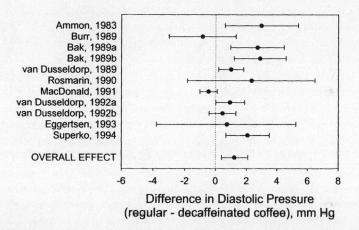

Figure 8.5. Meta-analysis by Jee and his colleagues of 11 experimental studies of the effects of caffeine on blood pressure. The dots show the average effects for each study; the lines are 95% confidence intervals. The bottom symbol in each panel is the overall average effect, with 95% confidence interval, for all studies. Modified with permission from Figure 1 in "The Effect of Chronic Coffee Drinking on Blood Pressure: A Meta-analysis of Controlled Clinical Trials," by S. H. Jee et al., *Hypertension* 33:647–652, copyright ©1999 by Lippincott Williams & Wilkins.

that included zero, suggesting that caffeine might not have an effect on blood pressure. How can we make sense of these divergent results?

Jee and his colleagues (1999) used a procedure called *meta-analysis* to help interpret the results of these various studies. The general concept of combining results of different studies in a quantitative way has a long history in statistics, dating back to at least 1904, but the modern version of the method was developed by Gene Glass in the 1970s to assess experimental evidence about the effectiveness of psychotherapy. Glass (1977) coined the name meta-analysis to indicate that this method was an analysis of analyses in order to integrate the results of various individual studies (Light and Pillemer 1984). The two basic objectives are to understand why different studies of the same question sometimes produce different answers and to see if a consensus can be achieved despite variation in the results of individual studies.

Notice that some of the confidence intervals for effects of caffeine on blood pressure in Figure 8.5 are very broad and others are much narrower. The Dutch study, labeled "van Dusseldorp, 1989" in this figure, has one of the narrowest confidence intervals of all. The widths are determined by the amount of variation among subjects in an experiment, as well as the total number of subjects. Figure 8.3 displays the variation for the 1989 Dutch study. If the histogram and corresponding normal curve in Figure 8.3 were more spread out, this would represent greater variation in response to caffeine among the 45 subjects and would mean a wider confidence interval for the average effect of caffeine. Likewise, if there were fewer subjects, the confidence interval would be wider. The widest confidence intervals in Figure 8.5 are for studies by Rosmarin and colleagues (1990), with 21 subjects, and Eggertsen and colleagues (1993), with 23 subjects; only one study (Ammon et al. 1983) had fewer subjects than these two.

The simplest way to combine the results of these 11 studies would be to add up the average effects of caffeine on systolic blood pressure for the individual studies and divide by 11 to get an overall average, then do the same for diastolic pressure. But this approach doesn't account for the fact that the studies differ greatly in the precision of their estimates of effects of caffeine on blood pressure. For example, the 1989 study by van Dusseldorp's group produced about the same average increase in systolic blood pressure with caffeine as the 1993 study by Eggertsen's group, but the confidence interval for the latter was much wider than that for the former, making the latter estimate of the average effect of caffeine less reliable (Figure 8.5). Therefore, meta-analysts generally calculate an overall effect size by using a *weighted* average of the mean effects from the individual studies. Studies with higher precision are weighted more strongly than less reliable studies. For systolic blood pressure, for example, the weighting factor for the 1989 van Dusseldorp study was about 0.24 and that for the 1993 Eggertsen study was about 0.004. If these were the only two studies, the estimate of the overall average effect of caffeine wouldn't be affected much by these different weighting factors because the means of the two studies were similar. But suppose the 1989 van Dusseldorp study, with a mean effect of $+1.5$ (caffeine increases systolic pressure), and the 1990 Rosmarin study, with a mean effect of -1.8 (caffeine decreases systolic pressure), were the only two available. The simple average of these two values is -0.15. However, the estimate of the mean for the van Dusseldorp study is more precise, so its weighting factor is larger than that for the Rosmarin study, 0.24 versus 0.01. Therefore, the weighted estimate of the average based on these two studies would be

$$\frac{0.24 \times (1.5) + 0.01 \times (-1.8)}{0.24 + 0.01} = 1.4.$$

This weighted average of 1.4 is much closer to the value of 1.5 from the van Dusseldorp study than to the value of -1.8 from the Rosmarin study and quite different from the simple average of -0.15. Appendix 2 discusses how these weighting factors are determined, together with some related statistical issues.

Based on their meta-analysis, Jee and his colleagues (1999) concluded that the overall effect of caffeine was to increase systolic blood pressure by about 2.4 units, with a 95% confidence interval of 1.0 to 3.7 units, and to increase diastolic pressure by about 1.2 units, with a 95% confidence interval of 0.4 to 2.1 units. These values define an objective consensus based on the combined results of the 11 individual studies. Because the 95% confidence intervals don't include zero, the meta-analysis provides reasonably convincing evidence that caffeine has a meaningful effect on blood pressure, although this effect is relatively small. Although nine of the individual studies showed greater systolic or diastolic pressure for subjects drinking caffeinated coffee than for control subjects, this effect was not significantly greater than zero in several cases. Therefore, meta-analysis helps to resolve some of the uncertainty arising from inconclusive or incompatible results of replicated tests of the same hypothesis.

Besides integrating results from various studies to reach a general conclusion, meta-analysts explore reasons for the differences in results. Jee's group found few factors associated with variation in the results of the 11 studies that they analyzed. For example, it didn't matter whether control subjects drank no coffee or decaffeinated coffee: the differences in systolic and diastolic pressure between regular coffee drinkers and control subjects were similar in both cases. Most of the other differences in experimental design also did not affect the results. Two factors that were important were the average age of the subjects and the amount of coffee consumed. Both systolic and diastolic pressure were more affected by caffeine in studies with younger subjects, and effects of caffeine on systolic pressure (but not diastolic) were greater in studies in which the subjects drank more coffee (the range for the 11 studies was 3 to 8 cups per day). The latter effect is particularly noteworthy because it implies a *dose-response relationship* between the amount of coffee consumed and systolic blood pressure: the greater the dose of coffee, the larger the response. Since physiological processes commonly show such patterns, evidence for a dose-response relationship between coffee consumption and blood pressure is consistent with fundamental mechanisms of physiology. Even without knowing the specific biochemical processes by which caffeine increases blood pressure, the existence of a dose-response relationship is stronger evidence for a biological effect than simply demonstrating that blood pressure is higher when caffeine is consumed than when it is not.

TIME AND THE EFFECT OF CAFFEINE ON BLOOD PRESSURE

The experiments discussed so far in this chapter lasted for 2 to 11 weeks, and the meta-analysis of these experiments by Jee and his colleagues (1999) showed that systolic pressure was about 2.4 units greater on average and diastolic pressure was about 1.2 units greater on average when subjects were drinking regular coffee than when they were drinking decaffeinated coffee or not drinking coffee at all. There have also been several experiments lasting less than 24 hours that showed more dramatic effects of caffeine on blood pressure. The reduced effect of caffeine after a few weeks illustrates a common physiological response called *acclimation:* our bodies adjust to an environmental condition (regular consumption of caffeinated coffee) by less extreme physiological changes (elevated blood pressure) as time goes on. What would this process of acclimation to caffeine look like over a period of years instead of weeks? Unfortunately, this question isn't amenable to experimental study because few people would volunteer to be randomly assigned to drink a specific amount of caffeinated coffee daily for several years or abstain from caffeinated coffee for several years. The only feasible way to address this question is to use a prospective study, in which individuals with different characteristics such as coffee-drinking habits are followed over time; a retrospective study, in which individuals with different degrees of hypertension are asked to recall their history of coffee drinking; or a similar nonexperimental approach. I introduced these types of studies in discussing the effects of vitamin C and other antioxidants on memory loss with aging in Chapter 2. Because subjects aren't randomly assigned to treatment groups, there are inherent ambiguities in interpreting the results of these types of studies that don't exist for a well-designed experiment. However, when an experiment is impossible, we have to rely on evidence from such a comparative observational study.

In 1947, Caroline Thomas began a long-term study of medical students at Johns Hopkins University that continues to provide a wealth of data on many aspects of health and disease. About 1,000 students who graduated between 1948 and 1964 volunteered for the study. Researchers collected data on the health and nutrition of these subjects while they were students; once the students became physicians, they filled out questionnaires periodically for up to 33 years. In addition to reporting their eating, drinking, and smoking habits, the participants measured their own blood pressure yearly. In 2002, a group led by Michael Klag (2002) analyzed the relationship between coffee consumption during and after medical school and blood pressure throughout the lives of these physicians. There were too few women (only 111) for statistical analysis, so all results relate to the 1,017 white males who supplied data beginning in medical school.

Eighty-two percent of the participants in this prospective study drank coffee regularly, and the coffee drinkers used an average of 2 cups per day. Each daily cup of coffee was associated with an increase of 0.21 units in systolic blood pressure and 0.26 units in diastolic pressure, averaged over the entire duration of the study. The effect on systolic pressure per daily cup of coffee was less than half of the effect seen in experiments lasting 2 to 11 weeks as summarized by Jee's group in their meta-analysis (0.52 units/cup), but the effect on diastolic pressure was similar to that in the experimental studies (0.25 units/cup). In analyzing the long-term data from the prospective study of medical students, however, Klag and his colleagues had to consider the possibility that other factors might be associated with coffee consumption and might influence blood pressure because the medical students weren't randomly assigned to drink specific amounts of coffee. For example, "The heaviest coffee drinkers [as medical students] tended to be slightly older than the men who drank less or no coffee [and] men who drank more coffee were more likely to drink alcohol and smoke cigarettes" (2002:657). These other factors could confound the apparent relationship between coffee consumption and blood pressure, but an analysis taking these factors into account produced very little change in the amount of increase in systolic and diastolic pressure that could be attributed to each daily cup of coffee.

The most important conclusions of this study related to the long-term effects of coffee drinking on the likelihood of hypertension at age 60. Hypertension, or high blood pressure, is associated with increased risk of heart attack, stroke, and other cardiovascular diseases. About 28% of participants in the Johns Hopkins study who drank coffee regularly during medical school were diagnosed with hypertension at age 60, compared to 19% of those who didn't drink coffee during medical school. However, when these percentages were adjusted to account for differences in smoking, alcohol use, exercise, degree of obesity, and family history of hypertension, the relationship between coffee and hypertension disappeared. This research by Klag's group illustrates the complexities of interpreting results of prospective and other purely observational studies, just as we saw for similar studies of antioxidants and aging in Chapter 2. The Johns Hopkins study also illustrates the limitations of relying solely on short-term experiments for understanding the full scope of biological processes. Physiological responses to caffeine appear to differ, depending on the time scale. At least for blood pressure, responses are strongest immediately after drinking a cup of regular coffee, weaker when comparing consumption of regular coffee to consumption of decaffeinated coffee or no coffee for several weeks, and weakest when considering a lifetime of coffee-drinking habits. Both rigorous experiments and careful analyses of long-term observational data were necessary to show this range of effects.

HETEROGENEOUS RESULTS IN MULTIPLE STUDIES

In addition to blood pressure, coffee drinking has been linked to several physiological processes that may affect health, and some of these links have been studied sufficiently to warrant meta-analyses of their own. There has been a great deal of interest in effects of coffee consumption on various types of cancer (Tavani and La Vecchia 2001). There is some evidence that coffee causes a modestly increased risk of bladder cancer, although this evidence is disputable because there are confounding variables in many of these studies. A large retrospective study published in 1981 suggested an association between coffee drinking and pancreatic cancer, but this association has not been confirmed in subsequent studies. One of the most interesting hypotheses has been that coffee actually protects against colorectal cancer. There are several mechanisms through which this might happen. One dietary factor that may contribute to colorectal cancer is compounds produced by cooking meat that induce mutations in the cells lining the digestive tract; coffee contains substances that may inhibit the detrimental effects of these compounds. Also, coffee causes a rapid increase in the activity of the colon, which causes a desire to defecate in some people; clearing the digestive tract frequently may reduce contact between the digestive lining and carcinogenic agents in food. Partly because of the possibility that coffee drinking might be beneficial for this aspect of digestive health, several researchers have studied relationships between coffee consumption and colorectal cancer. In 1998, Edward Giovannucci of Harvard Medical School reported results of a meta-analysis of these studies.

As you might guess, there have been no experimental studies of coffee and colorectal cancer. The nonexperimental research considered by Giovannucci fell into two categories: prospective studies and retrospective studies. The latter are also called case-control studies because a group of subjects with a disease (cases) is compared to a group of subjects without the disease (controls). The cases are often hospital patients; the controls may be other hospital patients with different diseases or healthy people from the general population. In either case, researchers usually attempt to match cases with controls in terms of age, sex, and other basic characteristics. Giovannucci found five prospective studies and 12 case-control studies suitable for his meta-analysis. Six of the case-control studies used population-based controls and six used hospital-based controls. Most of the studies that Giovannucci excluded from analysis simply didn't report results in quantitative form.

For each study that provided adequate data, Giovannucci classified subjects in two ways: those who consumed high or low amounts of coffee and those who were or were not afflicted with colorectal cancer. This relatively coarse classification of coffee consumption was necessary because the various studies reported the amount of coffee consumed in diverse ways. However, Giovannucci's "low" category represented 0 to 1 cup per day for most studies. This two-way classification leads to calculation of relative risks as a standard measure of the relationship between a factor such as coffee consumption and the likelihood of acquiring a disease. This term was introduced in Chapter 6 in the context of risks of cancer in twins. In the case of coffee and colorectal cancer, the relative risk of cancer as a function of coffee consumption is the proportion of individuals with cancer among those with high consumption divided by the proportion of individuals with cancer among those with low consumption. A ratio greater than 1 suggests that coffee increases the risk of colorectal cancer; a ratio less than 1 suggests that coffee protects against colorectal cancer.

Just as for the experimental studies of caffeine use and blood pressure, the precision of an estimate of relative risk can be represented by a 95% confidence interval. If this confidence interval is entirely above 1 or entirely below 1, there is justification for rejecting the null hypothesis that coffee has no effect on risk of colorectal cancer. In nonexperimental studies, however, this interpretation may be compromised by confounding variables that are correlated with a hypothetical risk factor. If these confounding factors actually cause the disease, a seemingly significant association between the hypothetical factor and the disease may be bogus. As discussed in relation to Klag's study of medical students at Johns Hopkins, coffee drinkers differ from non-drinkers in many ways that could influence risk of colorectal cancer, as well as hypertension. A partial solution to this problem is to adjust estimates of relative risk and their associated confidence intervals to take into account potentially confounding variables. Describing precisely how this is done is beyond the scope of this book, but the basic idea can be illustrated by the following example. The likelihood of colorectal cancer increases with age. Suppose older individuals are more likely to have a long history of heavy coffee drinking than younger individuals. Then the risk of colorectal cancer can be analyzed in relation to both age and coffee drinking to estimate the separate and independent effects of the two factors on risk. In principle, the same can be done for any set of explanatory variables that we think might influence risk of colorectal cancer. In practice, it's usually not feasible to measure, or even imagine, all of the variables that might be relevant, an idea that I introduced in Chapter 2. In the various studies of coffee and colorectal cancer analyzed by Giovannucci (1998), relative risks of cancer in response to high consumption of coffee were adjusted to account for confounding variables such as age, sex, diet, smoking, alcohol use, body weight, and so on, although different confounding factors were considered in the various studies.

The most interesting result of Giovannucci's analysis was that the overall relative risk of colorectal cancer differed for case-control studies and prospective studies. For the 12 case-control studies, the average relative risk of cancer in the group with high coffee consumption was 0.72, with a 95% confidence interval of 0.61 to 0.84. This means that people who drank relatively large amounts of coffee were about 72% as likely to get colorectal cancer as those who drank less coffee. Results for studies with hospital-based controls and population-based controls were similar. For the five prospective studies, however, the average relative risk in the group with high coffee consumption was 0.97, with a 95% confidence interval of 0.73 to 1.29. This implies no effect of coffee consumption on the risk of colorectal cancer. For the entire set of 17 studies, the average relative risk was 0.76 for high coffee drinkers, with a 95% confidence interval of 0.66 to 0.89. This overall average was similar to that for case-control studies alone because there were more case-control studies that included larger numbers of afflicted individuals, so they were weighted more heavily in the overall calculation.

Giovannucci interpreted these results as generally supportive of the hypothesis that coffee protects against colorectal cancer. He argued that the data from case-control studies were more credible because there were more cases of cancer in these studies and because their results were more consistent with each other than the results of the prospective studies. He also thought it was important that this consistency occurred in studies done in nine different countries on three continents. Giovannucci was fairly circumspect in his detailed report of this analysis in the *American Journal of Epidemiology:* "The results of this meta-analysis indicate a lower risk of colorectal cancer associated with substantial consumption of coffee, but they are inconclusive because of inconsistencies between case-control and prospective studies, the lack of control for important covariates in many of the studies, and the possibility that individuals at high risk of colorectal cancer avoid coffee consumption" (1998:1043). However, in a half-page summary of this work published in the journal *Gut*, Giovannucci dropped the caveats and simply stated, "Substantial coffee consumption was associated with a lower risk of colorectal cancer in the general population" (1999:597). Ekbom was even more enthusiastic in an accompanying commentary: "Whatever the biological mechanism, coffee drinkers are at lower risk of developing colorectal cancer compared with those who have chosen or been forced into a life without this stimulus" (1999:597).[2]

We should be cautious about accepting these conclusions at face value without thinking more carefully about the different results of the case-control and prospective studies. Although Giovannucci (1998) emphasized the case-control results, which suggested that coffee drinking might protect against colorectal cancer, he did so for the wrong reasons. He found more than twice as many case-control studies as prospective studies for his meta-analysis, but this was probably a simple consequence of the fact that case-control studies are easier to do, not because they are more reliable. Prospective studies require much more planning and funding than case-control studies because they necessitate a long-term commitment to follow their subjects. In addition, the larger number of diseased individuals typically included in case-control studies than in prospective studies does not imply that estimates of risk are more precise for the former, as Giovannucci assumed. In a prospective study, healthy individuals are selected to start with, and they are followed for many years until some become diseased; in a case-control study, diseased individuals are selected and compared with a control group. For relatively uncommon diseases, it's not surprising that the latter approach yields more cases than the former. But prospective studies often have very large total sample sizes, which can make risk estimates more precise even if the number of subjects who get a disease during the study is relatively small.

The most important difference between prospective and case-control studies is that there are greater opportunities for bias in the latter. The starting point for a case-control study is often hospital patients with a disease such as colorectal cancer. Because these studies are frequently done by researchers affiliated with medical schools, the cases are usually patients in large research hospitals, which is clearly not a random sample of people with the disease. Controls may be other patients in the same hospitals with different diseases; these individuals are obviously not representative of the general population, and the comparison of cases and controls may be compromised by bias in the selection of these subjects as controls. For example, a substantial fraction of hospital-based controls might have cardiovascular disease. If so, their physicians may have advised them to give up coffee. So the control group might have lower coffee consumption on average than the cases, but this wouldn't have anything to do with the absence of colorectal cancer in the

controls. Even when population-based controls are used, these are rarely selected randomly from the general population but rather for convenience in
collecting data. Case-control studies often depend on recalled information
from subjects about dietary or other habits from the past, sometimes many
years in the past, which provides opportunities for error. For these reasons,
the philosopher Ronald Giere wrote, "Retrospective [case-control] studies
may provide a good reason to undertake other studies, either experimental or
prospective. Nevertheless, evidence for causal hypotheses based on retrospective data alone cannot be regarded as being as good as evidence based on
equally well-executed experimental or prospective studies" (1997:238). Thus
Giovannucci's reliance on data from case-control studies to conclude that
coffee is beneficial for avoiding colorectal cancer seems misguided.

INCONCLUSIVE META-ANALYSES AND THE FILE DRAWER PROBLEM

At this point, we're left with a large question mark about the relationship between coffee and colorectal cancer. Implications of the case-control and
prospective studies summarized by Giovannucci are discordant. I find the
prospective studies, which suggest that coffee has no effect on colorectal cancer, more convincing because of their somewhat greater rigor than case-
control studies. Giovannucci gives more credit to the case-control studies,
which suggest a beneficial effect of coffee on the risk of colorectal cancer, although he does suggest that further prospective studies with large sample
sizes may help resolve the uncertainty. In fact, several additional studies of
both types were done in the 1990s (Piedbois and Buyse 2000). Most of these
studies were published after Giovannucci submitted his article to the *American Journal of Epidemiology* in July 1997. These new studies included a
prospective study of about 27,000 males in Finland (Hartman et al. 1998),
another prospective study of about 61,000 females in Sweden (Terry et al.
2001), and eight additional case-control studies. Figure 8.6 summarizes these
results. The average relative risks for the recent case-control and prospective
studies are 0.91 and 0.92, respectively. These are quite different from the average relative risk of 0.72 for the case-control studies analyzed by Giovannucci but similar to the average of 0.97 for the prospective studies analyzed
by Giovannucci. The confidence intervals for the new studies of both types
overlap 1.0, suggesting no significant effect of coffee consumption on colorectal cancer. Since the early case-control studies are the single discrepant
group of the four shown in Figure 8.6, the most reasonable inference is that
biases in these early studies may be responsible for their unusual results.
Therefore, we should tentatively accept the null hypothesis that coffee consumption has no beneficial or detrimental effect on colorectal cancer, although additional data might change the picture again.

This example shows how the conclusions drawn from a synthesis of research on a particular hypothesis may change as new studies are incorporated
in a meta-analysis. However, one of the biggest concerns about misleading
meta-analyses is not that new published studies may alter conclusions but

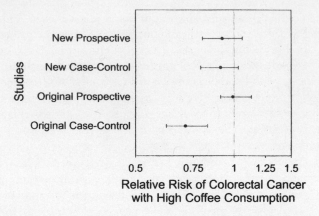

Relative Risk of Colorectal Cancer
with High Coffee Consumption

Figure 8.6. Meta-analysis of four sets of studies of the relationship between coffee consumption and risk of colorectal cancer. The original studies were those analyzed by Giovannucci in 1998. There were 12 case-control studies and five prospective studies, which contributed 20 and 13 estimates of relative risk, respectively, to Giovannucci's meta-analysis (some studies presented results for males and females separately, others for colon and rectal cancer separately). The new studies were published between 1992 and 2001 and provided 13 estimates of relative risk from case-control studies and four estimates from prospective studies (Hartman et al. 1998; Piedbois and Buyse 2000; Terry et al. 2001). The dot for each type of study is the overall average for that type; the horizontal lines show the 95% confidence intervals.

rather that unpublished studies might influence conclusions if they could be found and added to the meta-analysis. The main reason for this concern is the general belief that studies with dramatic or at least statistically significant results are more likely to be published than studies that don't produce evidence for rejecting a null hypothesis. Much research is done by graduate students and reported in their theses or dissertations. If they get negative results (e.g., no significant difference between a treatment and control group), they may not be strongly motivated to try to publish these results or they may be discouraged by their mentors from trying to do so. Journal editors may also be less likely to accept articles that report negative results. As a consequence, the literature readily available to a meta-analyst may be biased in showing significant effects. Therefore, a meta-analysis may produce a strong but incorrect conclusion if studies with negative results are not included. This is called the *file drawer problem* because a large proportion of studies with negative results may be sitting in filing cabinets in the offices of researchers, invisible to someone who wants to synthesize the literature through a meta-analysis (Sharpe 1997).

The file drawer problem is not a great concern for the example of coffee and colorectal cancer because the consensus from all published studies was that there was no relationship between these two factors (Figure 8.6). Unpublished studies, which are likely to confirm the lack of a relationship, wouldn't change this conclusion. However, when meta-analysis implies a significant result, it's important to think about how many unpublished studies of the problem might exist and whether including them might change the conclusion. Some meta-analysts go to great lengths to unearth unpublished studies: tracking down theses and dissertations, requesting reports from public health departments, and contacting colleagues who might know of studies that were never published.

CONCLUSIONS

This chapter introduced a smorgasbord of statistical methods that are important for interpreting research in biology and medicine. You may be somewhat disgruntled by the fact that we couldn't reach a very strong conclusion about either beneficial or detrimental effects of coffee despite suffering through these complex and abstract statistical analyses. Caffeine apparently has a small effect on blood pressure, but there is no solid evidence that a lifetime of coffee drinking increases the risk of hypertension when other factors that also affect hypertension are considered. Despite initial enthusiasm for the hypothesis, the general consensus from studies done through 2001 is that there is no significant relationship between coffee consumption and colorectal cancer. Nevertheless, the methods discussed in this chapter are important because they are used in medical and nutritional studies that are reported daily in the scientific literature and that are summarized frequently in the popular press. A news account may not use the term "meta-analysis" in describing a synthesis of individual research studies, but that is frequently what the story is about. The synthesis may involve experimental studies or prospective studies or case-control studies, each of which has strengths and limitations that were discussed in this chapter and elsewhere.

For some medical topics, meta-analyses of large numbers of studies have been quite decisive in influencing standards of care. For example, a group of researchers at Oxford University in Great Britain has done a series of meta-analyses of experiments to test the hypothesis that aspirin and compounds with similar physiological effects reduce the risk of heart attack and stroke. Many of these were long-term experiments with large sample sizes; the most recent meta-analysis, published in January 2002, summarized the results of 287 studies with more than 200,000 total patients (Antithrombotic Trialists' Collaboration 2002). The general conclusion of these analyses was that low doses of aspirin protect against various kinds of cardiovascular disease in high-risk patients; this conclusion has been widely accepted by practicing physicians, who prescribe aspirin to many of their patients.

For other topics, large-scale meta-analyses of contentious hypotheses have produced reasonably definitive conclusions, although these conclusions have not yet led to a paradigm shift among physicians. For several decades, physicians have recommended that patients reduce their intake of salt in order to reduce blood pressure and the risk of hypertension. The dogma that salt is detrimental for cardiovascular health was based on a variety of evidence, including experiments in which blood pressure was compared between people placed on low-salt diets or normal diets. Graudal and colleagues (1998) at the University of Copenhagen in Denmark reported a meta-analysis of 104 such experiments. About half of the experiments used subjects with hypertension; the other half used subjects with normal blood pressure. For individuals with hypertension, reducing salt intake apparently has a small but significant beneficial effect on blood pressure. For subjects with normal blood pressure, however, the meta-analysis of Graudal's group showed no average reduction in blood pressure from low-salt diets. The implications of this have been hotly debated by participants in the salt controversy, as described by Gary Taubes (1998), but widely ignored by physicians, who still recommend reducing salt intake to their patients regardless of whether they have hypertension or not.

The most interesting conclusion from these many experiments, however, is *not* the calculation of an average effect of salt reduction on blood pressure but rather the discovery that individuals seem to differ in sensitivity to salt. Most individuals have little or no change in blood pressure when they reduce salt intake, but some have a large decrease on the same low-salt diet. This produces a lot of variability in the results of experiments, which makes it difficult to detect a small but significant overall effect of salt on blood pressure. But this individual variation in salt sensitivity opens up new avenues of research, which may eventually lead to a clearer understanding of how salt affects our physiology and to more effective means of preventing and treating hypertension.

RECOMMENDED RESOURCES

Abelson, R. P. 1995. *Statistics as principled argument*. Erlbaum, Hillsdale, N.J. A refreshingly readable book about how statistical arguments are used in testing hypotheses, with lots of fascinating examples from the social sciences.

The Cochrane Collaboration. 2003. Preparing, maintaining and promoting the accessibility of systematic reviews of the effects of healthcare interventions. http:/www.cochrane.org (accessed October 1, 2003). The Cochrane Collaboration is an international organization that sponsors systematic reviews of a wide range of medical topics through meta-analysis. Their work is summarized here.

Hunt, M. 1997. *How science takes stock: The story of meta-analysis*. Russell Sage, New York. Although a bit too breezy and uncritical, this book describes many examples of meta-analysis and their implications for public policy.

Snell, J. L. 1999. Chance. http://www.dartmouth.edu/~chance (accessed October 1, 2003). Diverse source of materials for teaching and learning probability and statistics.

Weed, D. L. 2000. Interpreting epidemiological evidence: How meta-analysis and causal inference methods are related. *International Journal of Epidemiology* 29:387–390. Weed discusses the relationship of meta-analysis to other methods of testing hypotheses that we've considered in this book.

Chapter 9

How Will Climate Change Affect the Spread of Human Diseases?

Models and the Perils of Prediction

The *Los Angeles Times* reported in May 2002 that owners of ski resorts in Australia were buying new equipment to make snow worth $20 million in anticipation that global warming would cause reduced snowfall. This story indicates that a wide range of people has come to accept the fact that contemporary human activities affect global climate patterns. These people include not only virtually all scientists but also an increasing number of astute business leaders. Humans influence global climate in many ways, but the most familiar is the greenhouse effect. Earth is heated by solar radiation. Much of this is reflected back into space, but carbon dioxide and other compounds in the atmosphere absorb some of the solar radiation that has been reflected from Earth's surface. Thus some of the energy supplied by the Sun is retained in Earth's atmosphere. In fact, climatologists estimate that a hypothetical Earth without an atmosphere would be cooler by about 33 degrees Celsius (°C).[1] Burning fossil fuels releases carbon dioxide (CO_2) into the atmosphere, and the increased amount of CO_2 has been well documented (Figure 9.1). More carbon dioxide in the atmosphere implies greater heat retention, which causes increased temperatures at Earth's surface, as well as altered patterns of precipitation (Mahlman 1998).

There is abundant and diverse evidence that global climate change has already occurred in association with an increase of about 30% in atmospheric concentration of CO_2 since 1880. The average surface temperature of Earth has increased by about 0.6°C in the last century, and several of the warmest years on record occurred during the 1990s. The average sea level worldwide increased by 10 to 20 centimeters (4 to 8 inches) during the twentieth century. The thickness of ice sheets at the North Pole decreased by about 40%

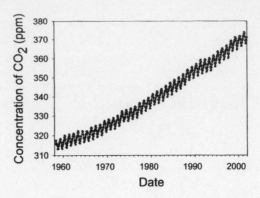

Figure 9.1. Atmospheric concentration of carbon dioxide in parts per million measured monthly at the Mauna Loa Observatory on Hawaii from 1958 through 2001. Researchers selected this site to monitor CO_2 because it is far removed from local sources of CO_2 emissions, such as power plants that are burning fossil fuels, so the data better reflect average atmospheric conditions in the northern hemisphere. The regular cycles illustrated in this figure represent increases in CO_2 during winter, when many plants are dormant, and decreases during summer, when they are using CO_2 in photosynthesis. An obvious increasing trend from year to year is superimposed on these seasonal cycles. Data from the Carbon Dioxide Information Analysis Center at Oak Ridge National Laboratory (http://cdiac.esd.ornl.gov/home.html).

between 1970 and 2000. Glaciers on tropical mountains are receding rapidly, and many are expected to disappear in the next 20 years. Droughts and severe storms have become more intense in recent decades. The average growing season in Europe has increased by about 11 days since 1960 (IPCC 2001; Johansen 2002). The biological effects of these changes in global climate are numerous and manifest, ranging from changes in the timing of reproduction of some species to altered distribution patterns and extinctions of others (see Chapter 4 for effects of global climate change on frog populations).

Carbon dioxide concentration in the atmosphere is increasing at a rate of about 1.5 parts per million per year. This seems like a minuscule amount, but if this rate of increase continues, the concentration in about 60 years will be twice that in 1750, near the beginning of the Industrial Revolution. Perhaps more telling, researchers have been able to estimate concentrations of CO_2 over long time periods by cutting cores from the thick ice sheets in Antarctica and Greenland and analyzing air bubbles trapped in the ice (IPCC 2001). This work has shown that *present* levels of CO_2 in the atmosphere are greater than any experienced on Earth in the past 420,000 years, and the rate of change in the last 100 years is unprecedented. Other evidence suggests that atmospheric levels of CO_2 may be greater now than at any time during the past 20 million years!

Although the general mechanisms of global climate change are well understood and some of the initial effects have been well documented, predicting future effects is much less certain and consequently more controversial. Scientific predictions about the future are usually derived from a model of some kind, so thinking about the structure and assumptions of models is important for evaluating their predictions. Of course, models and the process of prediction have much broader scope in science than simply trying to forecast future climate conditions. Some general discussion of these ideas will be helpful in setting the stage for more detailed discussion of the use of models in predicting impacts of increased CO_2 in the atmosphere.

PREDICTION

Whether associated with a formal model or not, prediction is a fundamental part of the scientific process because testing predictions is the basic means of evaluating hypotheses. An hypothesis may lead to predictions about how an experiment will turn out, so the predictions can be tested by performing the experiment to see if the results are consistent with the hypothesis. An hypothesis may also lead to testable predictions about new observations that can be made under natural conditions. For example, Steven Austad (1993) used an evolutionary hypothesis of senescence to predict that opossums on an island lacking predators would live longer and age more slowly than opossums on the nearby mainland. Since he made the prediction before knowing anything about the life span of opossums on the island, he could test it by initiating a study of these opossums (see Chapter 7). One pitfall of some observational studies in biology is the temptation to invent an hypothesis that is tested by "predicting" something that is already known and in fact was the basis for making the hypothesis in the first place. This isn't really prediction, but simply circular reasoning, and hypotheses built on this foundation have little credibility. For example, if Austad had *first* compared opossums on the mainland and island, and found that they had no predators and lived longer on the island, and *then* had come up with an evolutionary hypothesis to explain this relationship, he would have had to test the hypothesis in a different area or with a different species to avoid circular reasoning. It is often reasonable, however, to test an hypothesis by making a prediction about something that occurred in the past, as long as that event isn't known when the prediction is made. Paleontology is a very active area of current research, and new fossil discoveries often provide strong tests of hypotheses about processes of evolution. For example, scientists have hypothesized that whales and dolphins evolved from a particular group of hoofed terrestrial mammals, or ungulates, and several fossils found recently in Pakistan and nearby areas support this hypothesis because the fossils have a mixture of cetacean and ungulate characteristics (see Chapter 1).

Predicting the future can also be used to test hypotheses, but this has its own special difficulties. Unlike setting up an experiment or traveling to an island to study opossums in order to test an hypothesis, we have to wait to see if predictions about the future come true or not. Depending on the time scale of the predictions, it may take years or decades before questions about the hypothesis are resolved. For example, models of global climate change typically predict that average surface temperatures on Earth will increase by 1.5° to

4.5°C when CO_2 in the atmosphere reaches twice its preindustrial concentration, which is likely to occur before the end of the twenty-first century if current rates of fossil fuel use continue. This is a broad range of potential temperature changes, and climatologists expect that it will take several years to refine their models to make more precise predictions. Even when more precise predictions are possible, testing them by seeing what happens on the real Earth, as opposed to the virtual Earth represented in the models, will take a few decades.

This is a serious dilemma because policy decisions are linked to the predictions. The urgency of taking action to reduce CO_2 emissions depends on the validity of predictions about climate change, but these predictions can't be conclusively assessed for several years, at the earliest, by which time it may be too late to take effective action to avert the undesirable consequences of climate change. This uncertainty is frequently used as an excuse to justify inaction by some politicians and economists, especially in the United States, which is responsible for about 22% of global CO_2 emissions although it has only 5% of the world's total population (Johansen 2002). I'm not suggesting that this excuse is legitimate. In fact, a strong moral case can be made for the truly conservative position that, if there is a significant likelihood of more than minimal global climate change in the next century, we have a responsibility to our descendants to take action now to mitigate those changes.[2] But environmentalists like myself who take this position must accept the inherent uncertainties of trying to predict the future.

There is a more fundamental sense in which predicting the future is perilous. No prediction stands on its own, independent of an hypothesis or model. Unlike prophecies, which are absolute and unconditional, scientific predictions always take the following form: if a set of conditions is true, then something is predicted to happen. The conditions that lead to the prediction are encompassed in the hypothesis or model and may be relatively simple or extremely complex. If some of the conditions are false, the link between the hypothesis and the prediction is broken and the prediction isn't justified. Because of this link, predictions shouldn't really be stated without reference to the conditions of the hypothesis or model from which they are derived. For example, one prediction that underlies models of global climate change is that CO_2 concentration in the atmosphere will double from preindustrial levels by about 2060. This prediction is based on the assumption that current rates of emission of CO_2 into the atmosphere by the burning of fossil fuels and deforestation will continue for the next 60 years. It is also based on several assumptions about transfer of CO_2 from the atmosphere to the oceans and to terrestrial vegetation. These transfers account for about half of the total amount of CO_2 added to the atmosphere at current rates of fossil fuel burning and deforestation (i.e., the net increase of CO_2 in the atmosphere each year is only about half of what it would be if some of the added CO_2 wasn't taken up by ocean water and terrestrial plants; IPCC 2001). If rates of CO_2 emission change or if transfer processes among the atmosphere, oceans, and terrestrial vegetation change, the prediction that atmospheric CO_2 concentration will double by 2060 is no longer valid.

Unfortunately, predictions are often divorced from the conditions on which they are based, especially in the popular press. Sometimes this occurs because the conditions are very complex; other times the conditions are poorly understood, even by scientists. Nevertheless, it's always worth thinking critically about the assumptions that underlie any prediction and not simply relying on the claim of an expert authority that they are reasonable.

Thus predictions are always conditional on a set of assumptions. For this reason, predictions about the future always carry a degree of uncertainty, and the amount of uncertainty, increases as we try to predict farther and farther into the future. A different example may help clarify this unsettling situation. In recent years, various economists have predicted performance of the U.S. stock market for the next 10 to 20 years. These predictions range from the tripling of the Dow Jones industrial average to a market that gradually declines and remains well below its peak of the late 1990s for many years. These very different predictions are based on different assumptions about economic conditions, and there is no obvious reason why the most optimistic prediction is more or less plausible than the most pessimistic. Despite all the historical data that exist and despite the sophisticated modeling efforts of economists, future performance of the stock market is fundamentally unknowable. Thus all responsible financial advisors recommend that you employ a bet-hedging strategy and diversify your investments: put some money in the stock market in case it does very well, but put other money in bonds or other investments that may give an adequate return if the stock market does very poorly.

There is much greater consensus among climatologists about global climate change than among economists about future performance of the stock market, but predictions of climate in 50 to 100 years still carry some uncertainty because intervening conditions might change in unexpected ways. For example, emissions of CO_2 and other greenhouse gases would change dramatically following a global nuclear war or an epidemic that decimated the human population, but models of global climate change don't consider these possibilities. Even if atmospheric concentrations of CO_2 continue to increase at the current rate, a gradually changing climate could trigger a change in ocean currents, which might produce sudden and substantial *decreases* in worldwide temperatures. In fact, Kenton Taylor (1999) documented dramatic temperature changes in less than 20 years as recently as 12,000 years ago by studying ice cores from Greenland.

MODELS

Models lead to predictions that can be tested, but just what *are* models in science? The philosopher Ronald Giere (1997) defines a model broadly as a representation of the real world and uses maps as a basic illustration of models.

Many different kinds of models are used in science, ranging from verbal to pictorial to physical to quantitative models. Maps and diagrams are pictorial models; for example, Figure 9.2 shows a pictorial model of the global carbon cycle, illustrating both natural and anthropogenic (human-caused) transfers of carbon among the atmosphere, living organisms on land, and the oceans. One very important physical model in the history of biology was the scale model of DNA built by James Watson and Francis Crick in the early 1950s, which enabled them to elucidate the structure of the molecule that carries genetic information for all living organisms.

We've talked a lot about hypotheses and hypothesis testing in previous chapters, so we need to consider the relationship between models and hypotheses. Various authors define this relationship differently, but I think the most useful characterization is to say that a model is essentially a kind of hypothesis in which the assumptions are more explicit or the relationships are more specific or more detailed than in hypotheses not expressed as models. One important advantage of building models, especially quantitative models, is that it forces researchers to think more clearly about their assumptions. What assumptions are really necessary for a model and what assumptions can be dispensed with? How do predictions depend on changes in the assumptions? I don't want to imply that models are completely transparent—they can be compromised by hidden assumptions just like the most casual of hypotheses. But models often make the relationships between assumptions and predictions clearer, which promotes understanding.[3]

Models can be classified along several dimensions. In particular, quantitative models take many different forms. I'll consider four quantitative models to illustrate some of this variety, starting with the most complex and proceeding to the simplest. This strategy may seem perverse, but I'm not going to discuss the most complex type of model in detail because it involves chemistry and physics rather than biology. However, this general model of global climate change is the foundation for two of the biological models that follow, so I need to say a few things about it.

Models of global climate change are called general circulation models because they use equations to represent the circulation of substances throughout the atmosphere and explain the resulting heating and cooling of the atmosphere. As explained by J. D. Mahlman, "That early model [1965], as well as all of today's models, solves the equations of classical physics relevant for the atmosphere, ice, ocean, and land surface" (1998:89–90). Although based on fundamental equations, the models are very complex because the atmosphere is divided into a series of cells that cover the entire surface of Earth and it takes thousands of equations to represent processes occurring within and between cells. Current models divide the atmosphere into 10 to 30 layers over each block of 62,000 square kilometers of Earth's surface (about the size of West Virginia). Therefore, the models are typically explored by using the fastest supercomputers available. There are about 10 independently developed models in use by various research groups around the world; the summary assessments of the Intergovernmental Panel on Climate

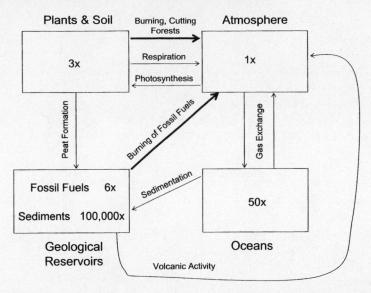

Figure 9.2. A schematic illustration of the global carbon cycle. The amounts of carbon in each compartment are expressed with reference to the amount in the atmosphere. For example, there is three times as much total carbon in soil and terrestrial plants (3x) as in the atmosphere (1x). The arrows indicate transfers of carbon between compartments. For example, plants use carbon dioxide in photosynthesis, so this process transfers carbon from the atmosphere to terrestrial plants. Two anthropogenic processes are contributing to the increased CO_2 concentration in the atmosphere illustrated in Figure 9.1: burning of fossil fuels and deforestation. These processes are shown by the bold arrows. See Reeburgh (1997) for more detailed information.

Change (IPCC) are based on these 10 models plus about 20 more that have been derived from them.

The models are typically refined by seeing how well their output corresponds to trends in temperature and other climatic variables that occurred during the twentieth century. Figure 9.3 illustrates this process. Despite year-to-year variation in average recorded temperatures, the output of the models based on increasing atmospheric CO_2 measured during the twentieth century match observed temperature trends fairly well.

In terms of future climate, IPCC (2001) considered several scenarios for emissions of CO_2 and other greenhouse gases during the twenty-first century to predict that average temperature in 2100 would probably be between 1.4° and 5.8°C higher than in 1990. Continued reliance on fossil fuels will lead to temperatures near the high end of this range; switching to renewable sources of energy will lead to temperatures near the low end (this is an oversimplification of the scenarios, but it includes their essential features). Andronova and Schlesinger (2001) recently did an analysis that suggested that the 90% confidence interval (see Chapter 8) for temperature increase by 2100 is wider than the range from 1.4° to 5.8°C considered likely by the IPCC in its 2001

Simulated Annual Global Mean Surface Temperatures

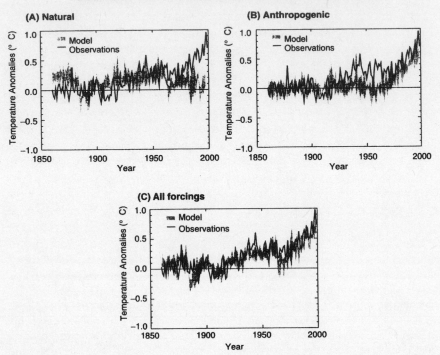

Figure 9.3. Consistency of a standard model of global climate change with temperature changes on Earth between 1860 and 2000. In all three panels, the dark lines are the recorded temperature data and the gray regions represent the predictions of a general circulation model. The model in panel (A) includes only the natural processes of annual variation in solar radiation and volcanic activity; the model in panel (B) includes only effects of human activity such as emission of greenhouse gases; the model in panel (C) includes both natural and anthropogenic processes and appears to fit the full range of data best. Zero on the vertical axis represents average global surface temperature between 1880 and 1920. Reprinted with permission from Figure 4 of *Climate Change 2001: The Scientific Basis*, by IPCC, copyright ©2001 by the Intergovernmental Panel on Climate Change.

report. According to Andronova and Schlesinger, the 90% confidence interval for temperature increase by 2100 is 1.0° to 9.3°C. Note that the extension is greater at the high end than at the low end and that an increase near 9°C (16°F) would make for an extremely hot world for our grandchildren.

General circulation models lead to a host of additional predictions about climate that are more alarming than this one. The rate of temperature increase is expected to be much greater than that which has already been seen in the twentieth century and will probably be greater than at any time in the past 10,000 years. Because the models work by dividing the atmosphere into cells above sections of Earth's surface, they predict regional differences in response to increased greenhouse gases. For example, temperatures in the northern parts of Asia and North America, especially in winter, are predicted to increase more than the overall global average of 1.4° to 5.8°C. There will probably be greater average precipitation worldwide, but some areas will be drier than they are at present and other areas will be much wetter. Perhaps more important, storms and droughts will probably be more intense and there will be greater year-to-year variation in precipitation than there is now. The average sea level will increase by 0.1 to 0.9 meters (4 to 35 inches) by 2100, inundating coastal areas worldwide. Ice caps and glaciers will continue to shrink and disappear. If the West Antarctic ice sheet collapses, as parts of the Larsen ice shelf did in 1995 and 2001, there may be a sudden and dramatic increase in sea level, which would have catastrophic consequences throughout the world (IPCC 2001; De Angelis and Skvarca 2003).

There is a high degree of consensus among climatologists and other researchers about these predictions of global climate change, for several reasons. First, the models are based on physical and chemical mechanisms that are very well understood. These mechanisms are supported by a large amount of experimental and observational evidence, which gives the models inherent plausibility. Second, climate changes that have already been observed in the twentieth century are consistent in detail with the models. Although "predicting" the known past isn't a strong test of a model compared with predicting the unknown future, the concordance between predictions of general circulation models and climate trends in the twentieth century supports the general hypothesis that changes in greenhouse gases influence climate. Third, the fact that several different models of future climate change produce similar predictions has contributed to agreement among scientists about the prospects of global climate change. The seemingly wide range of predictions for probable average global temperature in 2100 comes from exploring different scenarios for controlling emissions, not from applying different models to the same scenario. It's also worth noting that none of the 30+ models considered by the IPCC (2001) predicts that average temperature will be the same or lower in 2100 than today.

Despite the broad consensus about global climate change, there is still uncertainty about some aspects of climate models that climatologists will address in future work. Some of the most important concern the role of clouds, air pollution by particulate matter, and the oceans as a sink for carbon added to the atmosphere (Mahlman 1998). General circulation models predict increased water vapor in the atmosphere, which means more clouds. If these clouds are relatively low in the atmosphere, they would reflect solar radiation back into space and put a brake on global warming. If they are higher in the atmosphere, they would reflect infrared radiation from Earth back toward Earth's surface, acting as a positive feedback mechanism for global warming. The problem is that no one knows which of these possibilities is more likely. Similarly, sulfate particles released by burning fossil fuels mainly reflect solar radiation, whereas carbon particles released in fires mainly absorb solar radiation; these forms of air pollution would have opposite effects on global warning, but it's not clear which is dominant. Finally, the transfers of carbon between the atmosphere and the oceans aren't thoroughly understood, nor are the potential effects of global climate change on circulation patterns in the ocean. Both of these processes could have either positive or negative feedback effects on global warming.

Although models of global climate change are very complex, analyses of many of the possible biological effects of climate change are even more challenging. These include changes in the size and location of natural habitats, in behavior and physiology of organisms, and in abundance and distribution of species. They also include effects on agriculture and on human health. Some argue that agricultural effects of climate change will be beneficial. Because carbon dioxide is a basic resource for photosynthesis by plants, an increase in carbon dioxide should enable plants to grow faster, increasing crop yields. This argument is undoubtedly too simplistic because climate change may also benefit pests and weedy competitors of crop plants, so the net effect on agricultural productivity is unclear. The argument also ignores the prediction of regional differences in climate change. Some regions will experience more severe droughts, which will probably negate any potential contribution of increased CO_2 to greater crop yields.

Various authors have suggested a wide range of possible effects of global climate change on human health, from increased deaths due to heat stress to the adverse consequences of migration out of areas that become uninhabitable as temperatures increase (Martens 1999; Epstein 2000). The most dramatic effects will probably be associated with diseases that become more widespread in conjunction with environmental conditions favorable for disease-causing organisms. In particular, tropical diseases may expand beyond the tropics with warmer temperatures and greater precipitation in temperate areas. Malaria has evoked particular concern because it is responsible for so much morbidity and mortality, especially in Africa. In 1998, for example, there were an estimated 273 million cases of malaria worldwide, with 1 million deaths, mostly of children younger than 5 (Rogers and Randolph 2000). Because of the devastating effects of malaria in some parts of the world and because the biology and ecology of the disease are fairly well understood, researchers have developed models to predict the effects of global climate change on the distribution of malaria. Will malaria become more widespread with changing temperature and precipitation patterns across the globe? If so, what new regions will see an increase in malaria? Alternatively, could global climate change cause the range of malaria to contract? Two different types of models have been used to address these questions, which makes this a good example of contrasting approaches to modeling in biology, especially since the models make very different predictions.

A MECHANISTIC BIOLOGICAL MODEL OF THE GEOGRAPHY OF MALARIA

Before describing the first model, I should provide some background information about malaria. The disease is caused by a single-celled, protozoan parasite called *Plasmodium*. There are four species of *Plasmodium*, which produce different forms of malaria; cerebral malaria caused by *Plasmodium falci-parum* is the most damaging type and is responsible for most of the deaths attributed to malaria. The parasites may be transmitted to a person who is bitten by an infected mosquito. About 70 species of *Anopheles* mosquitoes are known to be vectors for the protozoans that cause malaria (Martens et al. 1999). Hundreds or thousands of *Plasmodium* may be transferred to a human host in one mosquito bite. In the human host, the parasites develop in the liver and then reenter the bloodstream and invade red blood cells, where cycles of asexual reproduction occur. During this process, the parasites destroy the red blood cells, causing many of the symptoms associated with malaria. Eventually, sexual forms of the parasite are produced and are taken up by mosquitoes, feeding on blood; the life cycle is completed when these sexual forms combine and produce a cell that develops into the stage that is transferred to people.

Both malaria parasites and mosquitoes are affected by temperature and moisture in various ways, so it's natural to try to predict the effects of global climate change on the distribution of malaria by linking predictions of climate models to the biology of the parasites and mosquitoes. A concept called the *basic reproduction rate* of the parasite provides a starting point for modeling the dynamics of malaria. The basic reproduction rate of a disease organism can be defined as the average number of secondary infections that occur following the infection of the first person in a population (May 1983). If the basic reproduction rate is greater than 1, the disease can spread within the population once it gets a foothold. Since we are interested in predicting whether changing climate will cause malaria to spread, the basic reproduction rate is a key variable.

What determines the basic reproduction rate of a parasite? George Macdonald (1961) introduced this concept in the 1950s and developed an expression for the basic reproduction rate of malaria that is still in use today. In fact, Pim Martens and several Dutch and British colleagues (1999) used Macdonald's formula as a foundation for modeling the effects of global climate change on the distribution of malaria.[4] This formula has three components: the number of mosquitoes to which the parasite is transferred when mosquitoes bite an infected person when the person is infectious, the probability that a mosquito lives long enough for the parasite to complete its incubation period in the mosquito, and the number of people to whom the parasite is transferred during the remaining lifetime of the infectious mosquito. The product of these three terms is the basic reproduction rate of the parasite, that is, the number of additional people that are likely to be infected once one person has malaria. Let's consider each of these terms individually.

The expected number of mosquitoes that will become infected from biting an infected person is the biting rate (number of people bitten by an average mosquito per day) times the number of mosquitoes per person in an area times the average duration of infectiousness of a person carrying malaria times the probability that the parasite is transferred successfully when a mosquito bites an infected person. For example, suppose each mosquito bites an average of 4 people per week, there are 10 mosquitoes per person, a person with malaria is infectious for 3 weeks, and the chance of a mosquito picking up the parasite when it feeds on a person is 25%. Then the expected number of mosquitoes that will become infected is $4 \times 10 \times 3 \times 0.25 = 30$.

The expected number of mosquitoes that will become infected if there is one infected person in a population depends on temperature because the biting rate is determined by how long it takes a mosquito to digest a blood meal, which is shorter at higher temperatures. For example, in one common species of mosquito there is no digestion below 10°C or above 40°C, and it takes 36.5 degree-days to digest one blood meal.[5] If the average temperature on a particular day is 30°C, this contributes $30 - 10 = 20$ degree-days toward the process of digestion of the meal (10 is subtracted from 30 because 10 degrees is the minimum temperature for digestion). If the temperature is 30°C on the next day, this contributes another 20 degree-days, for a total of 40, a bit more than necessary for full digestion. Thus we would expect a mosquito that fed at the beginning of a 30°C day to be ready to feed again near the end of the second such day. But suppose average temperature is 35°C on the two days. Then the amount of time necessary for full digestion is (36.5 degree-days)/(35 degrees $-$ 10 degrees) = 1.46 days, compared to 1.82 days at an average temperature of 30°C. In this example, the biting rate at an average temperature of 35°C would be 7/1.46 = 4.8 bites/week, whereas the biting rate at 30°C would be 3.8 bites/week.

The second element of basic reproduction rate is the probability that an infected mosquito will live long enough for the parasite to complete its incubation period. During the incubation period, the mosquito does not transfer parasites to people it bites because the parasites have not yet completed the portion of their life cycle that occurs in the mosquito. If the daily probability of survival of a mosquito is p and the incubation period of the parasite is T days, the probability that an infected mosquito survives long enough to become infectious to people is $p \times p \times p \ldots$ (for T days) $= p^T$. Both the daily survival probability of mosquitoes (p) and the length of the incubation period of parasites (T) depend on temperature. The former is less at higher temperatures but the parasites develop faster at higher temperatures, so the chance that the mosquito will live long enough to become infectious increases as temperature increases (Figure 9.4).

The final component of the basic reproduction rate is the number of people infected by one infectious mosquito during its remaining life after the parasite has completed its development. This equals the biting rate times the probability that parasites are transferred successfully from the mosquito to a person when the person is bitten times the life expectancy of the infectious mosquito. Biting rate increases with temperature as described above, but life expectancy decreases with temperature at the same rate because both depend on the length of time it takes a mosquito to digest a blood meal. Therefore, assuming that temperature doesn't affect the likelihood of the successful transfer of parasites from a mosquito to a person who is bitten, the third component of the basic reproduction rate is independent of temperature.

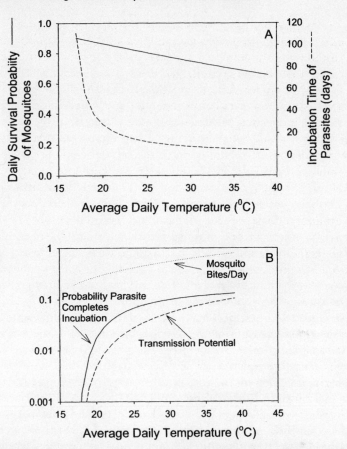

Figure 9.4. Key components of the biological model of the effects of temperature on transmission of malaria presented by Martens and his colleagues in 1999. Panel A shows the daily survival probability of mosquitoes (solid line, left axis, symbolized by p) and the incubation time of *Plasmodium falciparum* parasites in mosquitoes (dashed line, right axis, symbolized by T) in relation to average daily temperature. Panel B shows the probability that parasites complete their incubation (p^T) and the biting rate of mosquitoes in relation to temperature. The product of these two terms is the transmission potential, as discussed in note 6. Transmission potential increases by almost 80% with an increase in temperature from 20° to 21°C but by only 7% with an increase from 35° to 36°C. The lines are plotted from 17° to 39°C because parasites don't survive outside this range.

If you've followed the development of the model so far, you may be thinking about a much more important problem—the apparent assumption that moisture has nothing to do with the basic reproduction rate of malaria. It's common knowledge that mosquitoes need water to breed. Since global climate change includes precipitation patterns, as well as temperature, changes in precipitation may have a *big* effect on the distribution and abundance of mosquitoes and should be incorporated into the model.

In fact, Martens and his colleagues (1999) did consider precipitation in their model, although in a somewhat ad hoc way. African data suggested that rainfall of at least 80 millimeters per month (about 3 inches) is required for malaria to be continuously present, so Martens's group assumed that 80 millimeters of rainfall per month for 4 successive months was necessary for seasonal transmission of malaria. Since contemporary models of global climate change predict monthly precipitation, as well as temperature, for various regions of the world, Martens's group used two criteria to predict the risk of malaria in each specific region under future climatic conditions: (1) if predicted rainfall was greater than 80 millimeters for 4 or more successive months, they based the estimate of transmission on predicted temperature in the region; (2) if there was no sequence of 4 months with predicted rainfall greater than 80 millimeters, they assumed that there was no possibility of the transmission of malaria regardless of temperature. Keep in mind that one prediction of general circulation models is that some parts of the world, especially in the tropics, will become drier in the twenty-first century, which might lead to shrinkage of the range of malaria if the dryness causes precipitation to fall below the threshold for mosquito reproduction.

One additional assumption made by Martens and his colleagues was that the current global distribution of *Anopheles* mosquitoes would not change in response to global climate change. At present, mosquitoes that are capable of carrying malaria are much more widely distributed than the disease itself. For example, *Anopheles* mosquitoes occur in most of the United States, all of Europe, and much of Russia, although malaria is currently sporadic in these areas. Finally, the researchers had no data indicating how temperature might affect the abundance of mosquitoes, so they assumed that there would be no relationship between temperature and abundance. They considered these assumptions to be conservative. In other words, they thought that if they had not had to make these assumptions, the predicted effects of climate change on malaria would be even greater than with the assumptions. Why would this happen? There is good evidence that increased temperature causes faster reproduction in mosquitoes, which should increase their abundance. Unfortunately, this can't be effectively modeled at a global scale because of inadequate data on how population growth of mosquitoes responds to temperature, so Martens' group assumed no relationship between temperature and abundance in their model. As we saw in Chapter 6, all models in science rest on assumptions. The key to using models successfully is not to try to eliminate all assumptions, which is futile, but to make assumptions explicit and think carefully about how they may influence the results.

Martens and his colleagues began the evaluation of their model by examining its consistency with the current worldwide distribution of malaria. Large areas in eastern North America, Europe, Central Asia, and China have climate conditions that are suitable for malaria, although they don't have significant problems with malaria now. Indeed, malaria was present in these areas historically but was largely eliminated after World War II by mosquito control and other public health measures. In a few areas, notably in the western United States and western China, *Anopheles* mosquitoes are present but climate conditions are unsuitable for malaria parasites. Finally, very small areas on the Arabian Peninsula harbor malaria, although the model suggests that they shouldn't. The reason for this discrepancy is that monthly rainfall is less than 80 millimeters; evidently mosquitoes don't require quite this much precipitation to sustain viable populations.

Martens and his colleagues used five scenarios of climate change based on two different versions of a British general circulation model to predict the effects of climates expected in 2020, 2050, and 2080 on the distribution of malaria. The results for the five scenarios were generally similar, so I'll discuss only those for the scenario based on the most current and most detailed climate model. The results were expressed in terms of a relative measure of the risk of malaria called *transmission potential*.[6] Except for a few tiny areas (e.g., in Namibia in the southern part of Africa), the transmission potential was greater under future climate conditions than under current conditions. Transmission potential increased the most in temperate areas of North America and Europe because "an increase of, for example, 1°C around the minimum transmission temperature causes a larger increase of the transmission potential than the same increase at higher temperatures" (Figure 9.4; Martens et al. 1999:S99). The researchers also considered the likely effects of climate change on the length of time each year that people would be vulnerable to malaria. Areas such as the United States and Europe would probably experience epidemics or seasonal transmission, whereas year-round transmission would primarily occur in tropical areas. However, the model suggests that in Africa the area at risk of year-round transmission would decrease and the area at risk of seasonal transmission would increase from current conditions. The reasons are that climate models predict that precipitation will decrease in some parts of Africa and that mosquitoes require adequate moisture to reproduce. Unfortunately, this isn't necessarily a reason for optimism because individuals can develop and maintain partial immunity to malaria when there is continuous transmission. Therefore, populations may experience more adverse effects from regular seasonal transmission of malaria than from year-round transmission.

Martens and his colleagues linked the results of their model of the geography of malaria to predictions of population growth in various regions of the world to conclude that the number of people at risk of acquiring malaria would increase by about 20% between now and 2080. Two-thirds of this increase, amounting to 300 million people, would be at increased risk of cerebral malaria, which is responsible for most deaths. Many of these additional people would live in North America or Europe, and a skeptic might argue that malaria could be kept under successful control in these developed countries. However, social changes are even more difficult to predict than biological and climatological changes, so assuming that the public health infrastructure in North America and Europe will remain as effective in 2080 as it is today is risky. Indeed, malaria has recently become endemic in three southern Asian countries (Azerbaijan, Tajikistan, and Turkey), where the public health system has broken down. The researchers summarize their response to our hypothetical skeptic as follows:

> With planning and development of adaptation capacity, increases in disease incidence associated with climate change may be largely prevented. However, many of the current technical, socio-economic, and political barriers to successful prevention and control will also apply in the future. In most poor developing countries, the malaria burden of climate change will therefore be left unmitigated or be mitigated at high cost to the economy. In either case, the sustainability of human health and well being would be in greater jeopardy as without a human induced climate change. (Martens et al. 1999:S105)

In recent decades, malaria has become more common at high elevations in Africa, Latin America, and some parts of Asia. Paul Epstein (2001) has interpreted this development as evidence that global climate change is already increasing the risk of tropical diseases because the increased incidence of malaria at elevations above 1,500 meters (5,000 feet) parallels other climate-related changes on tropical mountains, such as the shrinkage of glaciers and the extension of the ranges of some plants to higher elevations. This interpretation has been challenged, however, because several other changes have also occurred that could be responsible for the expansion of malaria into tropical highland areas: immigration of people into the highlands; deforestation, which creates a more suitable habitat for mosquitoes; social upheaval, leading to the deterioration of public health measures such as mosquito control; and increased resistance of parasites to antimalarial drugs.

The best evidence for an association between climate changes that have already occurred and increased incidence of malaria comes from a study in the Northwest Frontier Province of Pakistan by M. J. Bouma and colleagues (1996). This part of Pakistan is among the most northerly areas in the world subject to cerebral malaria. During the 1980s, the number of cases increased from fewer than 1,000 to more than 25,000 per year. Bouma's group analyzed the relationship between the incidence of cerebral malaria and several weather variables for the period from 1978 through 1993 because mosquito control procedures were relatively consistent during this time. Thus, the results would not be confounded by changes in control operations. Cerebral malaria is most common in the fall in northwestern Pakistan, and Bouma's group found that warmer and wetter falls were associated with more cases of cerebral malaria.

A DESCRIPTIVE STATISTICAL MODEL OF THE GEOGRAPHY OF MALARIA

David Rogers and Sarah Randolph of the Department of Zoology at Oxford University took a very different approach in modeling the effects of global climate change on the distribution of malaria than the biological model developed by Martens and his colleagues. Rogers and Randolph (2000) argued that not enough information was available to justify the biological model. In particular, they were skeptical of the fact that Martens's group didn't consider the effects of climate on the numerical abundance of mosquitoes but only a relationship between rainfall and the presence or absence of mosquitoes. Although Rogers and Randolph concluded that use of the biological model was premature, they thought that eventual construction of a robust biological model would be a worthy goal.

As an alternative, they developed a purely descriptive, statistical model, using information on current climatic conditions in areas with and without malaria to determine the critical climatic factors associated with the disease. Then they used the relationships derived from this analysis to predict the distribution of malaria in 2050. This approach ignores the specific mechanisms by which climate influences mosquitoes or parasites, which are the foundation of the biological model. Instead, it simply asks if there are strong correlations between particular climatic factors and the presence of malaria, without considering why these correlations might exist. The value of this statistical approach is judged purely by the success of its predictions, not by its contribution to understanding how things work. As we will see, Rogers and Randolph were able to test their model even though 2050 is several decades in the future.

They began their analysis by randomly choosing 1,500 locations within the current range of cerebral malaria and 1,500 locations outside this range but close enough that climate conditions weren't obviously unsuitable for the disease. They then gathered published data on temperature and moisture for each of these 3,000 locations, using a total of 15 climatic variables in their model. These variables represented both average and extreme conditions, either of which might influence the distribution of malaria.

Rogers and Randolph found that five of these variables best differentiated the locations with and without malaria. As you might have guessed, drier and colder sites were less likely to have malaria than wetter and warmer sites. But some subtleties in the results were noteworthy. For example, sites lacking malaria differed from sites with malaria primarily in *minimum* monthly precipitation, not *average* monthly precipitation. This suggests that malaria parasites or their mosquito carriers depend on a threshold level of precipitation each month rather than a large total amount of precipitation per year. If there was a large yearly total but pronounced wet and dry seasons, the minimum monthly precipitation during the dry season might be below the threshold to support malaria.

Rogers and Randolph tested their model by randomly selecting 20,000 locations on Earth's surface and using an equation containing the five climatic variables to calculate the probability that malaria was present at each of these locations, which were different from the 3,000 that were used in the statistical analysis that produced this equation. If the probability of malaria at a site was greater than 0.5, they classified the site as a likely malaria site; otherwise they classified it as a likely nonmalaria site. They then compared this classification with the actual status of the sites. Overall, this process classified 78%

of the sites correctly. About two-thirds of the errors were false positives, sites without malaria that should have had malaria according to the model. The remaining errors were false negatives.

In summary, this statistical model enabled Rogers and Randolph to predict where malaria should occur, based on five key climatic variables. They were able to test the model against the current distribution of malaria by using one portion of the data to develop the model and another portion (i.e., different locations) to test the model. They used a similar procedure to compare the predictive ability of their model to that of the biological model of Martens's group. Rogers and Randolph reported that the percentage of correct classifications of sites with and without malaria based on the biological model was 67%, compared to 78% for their statistical model.

Finally, they linked the predictions of global climate in 2050 to their statistical model of the distribution of malaria to predict how much the range of malaria would increase by 2050. Their results suggested "only a relatively small extension . . . compared to the present-day situation: northward into the southern United States and into Turkey, Turkmenistan, and Uzbekistan; southward in Brazil; and westward in China. In other areas, malaria was predicted to diminish" (Rogers and Randolph 2000:1,765). In contrast to the conclusion of Martens's group that 300 million additional people might be at risk of cerebral malaria because of global climate change in the twenty-first century, Rogers and Randolph suggested that increased risk would affect at most 23 million people.

PREDICTION AND UNDERSTANDING

How should we evaluate the relative merits of these two models? If, for the sake of argument, we accept the claim by Rogers and Randolph that their statistical model is more consistent with the current distribution of malaria than the biological model developed by Martens's group, perhaps we should have more faith in the predictions of the statistical model. At least for malaria, these predictions about the impact of global climate change on the number of people at risk are much less dire than those of the biological model. However, models in science aren't developed simply to make predictions but also to help increase our understanding of how the world works. In fact, making (and testing) predictions is a tool for improving understanding, not usually an end in itself. With this is mind, we should take a second look at the two types of models.

Both the statistical and biological approaches to modeling provide a framework for improved understanding of the global distribution of malaria and other diseases. The main contribution of the statistical approach is to emphasize the fact that the presence of a disease in an area may be influenced by multiple, interacting climatic factors. For example, five variables were important in distinguishing areas with and without malaria, and three of these variables described threshold conditions rather than average conditions, suggesting that a model based solely on average temperature would be too simplistic. In addition, a relatively sophisticated statistical model like that of Rogers and Randolph incorporates interactions between the variables used to predict the presence and absence of malaria. For example, it might be that less precipitation was required to support malaria in a warmer region than in a cooler region, contrary to the assumption that the disease is more successful as the weather becomes warmer and wetter. We discussed the significance of interactions between genetic and environmental contributions to cancer in Chapter 6; the malaria example illustrates the general importance of thinking about interactions in trying to understand biological phenomena.

The major limitation of the purely statistical approach is that it ignores the mechanisms by which climate and other factors affect the population dynamics of the disease. By contrast, these mechanisms are the heart of the biological approach. Therefore, the biological approach is concerned with understanding why a disease is more prevalent under some conditions than others, not simply in describing those conditions. In short, a biological model is a model of causation, whereas a statistical model is a description of a pattern of correlations. In the case of malaria, the statistical model of Rogers and Randolph was essentially just a description of correlations between climatic variables and the presence of malaria. Granted, this description was more complex than the simple correlations between memory ability and levels of antioxidants in the blood, discussed in Chapter 2, but the basic approach used in the two cases is similar.

By contrast, the biological model of malaria was based on a set of explicit assumptions about how temperature affects the life cycles and behavior of mosquitoes and parasites. Because the model was based on these mechanistic assumptions, it may stimulate experimentalists and field researchers to collect better data about relationships embodied in the assumptions. The model may also stimulate theoreticians to explore its sensitivity to changes in the assumptions. Finally, a mechanistic, biological model of a disease can provide a rationale for testing various public health strategies based on manipulating key factors that affect the biology of disease organisms. Descriptive, statistical models don't have such obvious practical applications.

SOME PRACTICAL AND ETHICAL IMPLICATIONS OF BIOLOGICAL MODELS OF DISEASE

The contributions of biologically based models to understanding disease processes can be illustrated by elaborating and generalizing the concept of basic reproduction rate. As noted earlier, the basic reproduction rate is defined as the number of additional people that become infected following infection of the first person in a population. This indicator of the likelihood that an infection will spread in a population applies to diseases in general, not just malaria, although the specific formula depends on how the disease is transmitted. Let's imagine that some individuals have been vaccinated, so they are immune to the disease. If the basic reproduction rate in the absence of vaccination is represented by R_o and the proportion of the population immunized is represented by p, the basic reproduction rate of the parasite in this population is reduced to $R_o(1 - p)$ because there are fewer people susceptible to the disease. Consider a disease like measles, for example, which is transmitted directly from person to person. The basic reproduction rate of measles is determined by the number of virus particles shed by a person while the person is infectious, the number of susceptible people he or she comes in contact with during this time, and the probability of successful transmission for each of these contacts. If no one has been vaccinated and everyone is susceptible, the basic reproduction rate is at its maximum value of R_o. If half of the people contacted by the initially susceptible person have been vaccinated, the virus is not spread successfully during these contacts and the basic reproduction rate is at half of the maximum value

For a disease to spread in a population, the basic reproduction rate must be greater than 1:

$$R_o (1 - p) > 1$$

As the fraction of the population immunized (p) increases, the lefthand side of the equation, representing the basic reproduction rate, decreases. When p is large enough, the basic reproduction rate drops below 1 and the disease should not be able to spread. The equation can be rearranged to express the proportion of the population that must be immunized for the basic reproduction rate of the disease to be less than 1:

$$p > 1 - 1/R_o$$

Robert May (1983) estimated that R_o was about 80 for cerebral malaria in northern Nigeria in the 1970s. This implies that, if there were a successful vaccine for malaria, at least 99% of the population would have to be vaccinated to eliminate the disease from a country like Nigeria, where conditions for transmission are so favorable ($1 - 1/80 = 98.8\%$). Based on this model, a more effective strategy for combating malaria would be to try to reduce R_o by controlling mosquito populations; and in fact, this has been a major emphasis of public health programs in areas where malaria is prevalent. By contrast, the basic reproduction rate of smallpox before it was eradicated was about 4, which meant that at least 75% of a population would need to have been vaccinated to eliminate smallpox from a region. A key reason for the success of the worldwide campaign to eliminate smallpox was the relatively low value of R_o for this virus, which made it possible to drive the reproduction rate below 1 by vaccinating a substantial fraction of people without having to vaccinate almost everyone.

This model has some significant implications in relation to recent suggestions that mass vaccination programs for smallpox be started in the United States and elsewhere. Although smallpox was eradicated in human populations in the 1970s, the United States and then Soviet Union maintained stocks of the virus, ostensibly for further research. With the increasing boldness of terrorist organizations in 2001 and 2002, people became concerned that these groups might have access to smallpox and would be able to use it in a devastating attack. Should public health agencies make smallpox vaccine available to anyone who requests it? Should they require large numbers of people to be vaccinated to prevent such a terrorist attack from being successful? One of the costs of this strategy is that the vaccine itself has a risk of serious side effects or even death. Alex Kemper and his colleagues (2002) at the University of Michigan estimated that there would be about 285 deaths and 4,600 cases of severe side effects of vaccination if 178 million Americans were vaccinated against smallpox.

If a voluntary program of vaccination for smallpox were instituted, would you participate? Because of its relatively low basic reproduction rate, smallpox shouldn't be able to spread in a population if at least 75% of people are immunized. This means that if enough people participate in an immunization program, the sensible strategy for you as an individual is *not* to be immunized, to avoid the small but real possibility of complications or even death from the vaccine. Unfortunately, if this strategy is adopted by more than 25% of people, it won't work in checking the spread of smallpox. This is another example of the paradox of prevention introduced in Chapter 8: what makes sense for an individual is not necessarily optimal for a population.[7]

We've covered a lot of ground in this chapter, from global climate change to vaccination. My main objective has been to illustrate the role of quantitative models in the scientific process. We considered four models, beginning with a general circulation model of climate change that requires a supercomputer for analysis and ending with a simple algebraic model of how broad a vaccination program must be to control a disease. The two major models discussed in this chapter were a biological and a statistical model of the effects of climate on the geography of malaria, which differed in their relative emphasis on understanding causal mechanisms versus accurate prediction of the current distribution of malaria. As with some examples in other chapters, we didn't reach a definitive conclusion about the credibility of the very different predictions of these two models. However, I argued that the biological model may be more useful because it suggests lines of further research, both basic and applied.

Another important thread connects global climate change with vaccination. I used a model to suggest that what's best for a population, vaccinating as many people as possible, may not be best for an individual if enough people are vaccinated to reduce the basic reproduction rate of a disease below 1. This is analogous to a famous ecological principle that Garrett Hardin (1968) called "the tragedy of the commons." In an earlier era, villages had common grazing areas. If one family added one cow to its herd, that family got all the economic benefits, but the costs, in terms of degradation of the commons, were shared by all. Therefore, a rational cost-benefit analysis would convince each family to keep adding to its herd. The eventual result, however, would be catastrophic overgrazing of the commons so that it couldn't support any livestock. Today, the atmosphere is such a commons, and our use of fossil fuels to support our standard of living is the same as adding extra cows to the common grazing area of a nineteenth-century village. We benefit individually by our reckless use of resources, but in so doing we contribute to runaway global warming, whose numerous costs will be shared by all of our descendants. Just as with vaccination, what's good for the individual is not necessarily good for the population. How should we deal with this dilemma?

RECOMMENDED READING

Cohen, J. E. 1995. *How many people can the Earth support?* Norton, New York. A thorough and erudite discussion of the past, present, and future of human population growth, including an excellent exposition of some key strategies for interpreting different kinds of models.

Hardin, G. 1968. The tragedy of the commons. *Science* 162:1243–1248. A classic reference in environmental biology.

Johansen, B. E. 2002. *The global warming desk reference.* Greenwood Press, Westport, Conn. Complete summary of the climatology, biology, economics, and politics of global climate change.

Rose, G. 1992. *The strategy of preventive medicine.* Oxford University Press, Oxford. Very readable introduction to the paradox of prevention, discussed in Chapters 8 and 9.

Chapter 10

Conclusion
How Science Works and Its Role in Society

In the preceding chapters I used several examples from biology and medicine to explore some important aspects of the scientific process. Some of these examples have been prominently featured in recent news reports (leg deformities and declining populations of frogs; genetic and environmental contributions to the risk of cancer); others have been interesting mainly to specialists (spatial memory abilities of food-storing animals). Many of the examples have practical significance, although I selected them primarily for what they could teach about the scientific process rather than for their importance in affecting personal or social decisions, such as the use of caffeine (Chapter 8) or attempts to limit emissions of greenhouse gases (Chapter 9). I purposely used a diversity of examples, partly because I enjoyed the opportunity to learn about a broad range of topics and partly in hopes of stimulating your curiosity about something new. For example, you may have been attracted to the medical examples when you started reading the book but become intrigued by the stories of police dogs or troubled frogs or birds with prodigious memory abilities.

Although I used these stories to illustrate some general principles about how science works, I hope you found some of the biological details fascinating. Certainly basic principles of research design, statistics, and the philosophy of science are important, but they can be difficult to appreciate if discussed in a vacuum. Therefore, in addition to my goal of bringing some of these principles to life for readers lacking a scientific background, I simply wanted to tell a few exciting stories about discoveries in biology and medicine, focusing on the discovery process rather than the end results.

In fact, one of the key things that unites these various stories of work in progress is that none of them really had final conclusions. Therefore, it made sense to concentrate on their methods, although I didn't purposely pick examples that were inconclusive just to make it easier to make my point. In their incompleteness, these examples are representative of science in general. Some of my stories included more real progress in understanding than others, but even the former raised new questions for further research. For example, Vander Wall and others clearly showed that various birds and mammals that store food have impressive abilities to remember storage locations (Chapter 5). But this conclusion led to new questions about the brain structures involved in spatial memory and ongoing comparative studies of the brains of species with different food-storing behaviors. In essence, good stories in science are like good novels, in which loose ends are *not* all resolved by the last page but rather the pleasure of thinking about ideas extends beyond finishing the book.

Most of my examples were from ecology and medicine, two fields in which it is often especially difficult to bring closure to important questions. One reason for this difficulty is the great diversity of entities that ecologists and medical scientists study. Think of the many different types of habitats that exist on Earth, from arctic tundra, with its permanently frozen subsoil and brief growing season, to the Atacama Desert in Chile, which may not get any rain for years in succession, to deep-sea vents, which harbor life forms that thrive in near-boiling water without oxygen or light. At least 1.5 million species of plants, animals, and microbes occupy the multitude of habitats on Earth; these species are the primary objects of study by ecologists.[1] The variety of living conditions and lifestyles of even a tiny subset of these species, such as the amphibians discussed in Chapter 4, makes it unlikely that a single explanation exists for such phenomena as declining populations of the species worldwide.

Medical scientists face the same challenges as ecologists in confronting enormous diversity in their subjects of study, individual human beings. All 6 billion of us (except those few who have identical twins) are genetically unique, and our varied environments have influenced us in countless different ways. Although the understanding of basic biochemical and physiological processes of human beings can advance fairly steadily despite our diversity as individuals, answering questions about health and nutrition is much more challenging, as illustrated in Chapter 2 on antioxidants, Chapter 6 on cancer, and Chapter 8 on coffee. Individual variation among people because of genetic and environmental differences produces different responses to substances like vitamin C and caffeine, even under controlled experimental conditions. This variation also means that the causes of individual cases of cancer and other diseases are diverse and complex. In fact, the causes and consequences of individual variation are just beginning to be seen as important and interesting questions for research, not only in humans but also in other animals.

Individual variation among humans and members of other species and differences among the large number of species that exist make it difficult to get definitive answers to questions about the health and disease of humans, the behavior of animals, and the adaptations of species to changing environments. However, these questions are often those for which our personal needs or our social institutions demand clear and convincing answers. For example, consider the use of hormone replacement therapy by postmenopausal women in industrialized societies. Menopause at age 50 or so brings changes associated with reduced levels of estrogen and progesterone, two hormones that are produced by the ovaries in regular cycles until menopause. These reduced levels of sex hormones in older women have many effects, ranging from uncomfortable hot flashes to osteoporosis, which can contribute to life-threatening fractures. Therefore, it has become popular in Western society to prescribe hormone replacement therapy (HRT) for postmenopausal women, and many of these women take daily doses of estrogen or estrogen plus progesterone. Additional benefits of HRT besides alleviating symptoms of menopause and reducing the risk of osteoporosis have been widely touted. For example, comparative observational studies of women who used HRT and those who didn't suggested that HRT might protect against heart attacks and stroke. These studies were not really conclusive, however, because women who chose to use HRT were probably more health conscious than those who didn't—eating better, exercising more, and visiting their doctors more often. The reduced incidence of cardiovascular disease in the HRT group could have been due to any of these other factors rather than to HRT. Although this limitation of the early observational studies wasn't widely appreciated, it did lead to the organization of two large, randomized, clinical trials of HRT.

In the summer of 2002 physicians and their female patients were shocked to learn the initial results of these experiments, which were just the opposite of the observational data: HRT actually increased the incidence of cardiovascular disease. The experiments were stopped, but women approaching and past menopause were left in limbo (Enserink 2002). Analysis of the experimental results suggested that among 10,000 women using HRT, there would be about 7 more heart attacks, 8 more strokes, and 8 more cases of breast cancer per year but 6 fewer cases of colorectal cancer and 5 fewer hip fractures. These are statistically significant effects, meaning that they are unlikely to be due to chance, but what is their biological significance? Regardless of HRT, there is great variation among individual women in severity of hot flashes, vaginal dryness, degree of osteoporosis, and other consequences of menopause. There is probably a similarly large variation in individual responses to HRT. For some women, its benefits may greatly exceed risks; for others, the reverse may be true. The problem is that no one knows how to predict which individuals will benefit from HRT and which will not. Therefore, each woman's choice about using hormone replacement therapy is not at all obvious.

Just as individuals have to make personal decisions in the face of tentative and uncertain answers provided by science, legal and political institutions face similar dilemmas. These problems are exacerbated not only by widespread misunderstanding of the role of science in deciding issues of importance to society but also by the basic workings of law and politics in our society. How do they differ from science in a way that can result in the misuse of science by these social institutions?

THE UNEASY INTERSECTION OF SCIENCE, LAW, AND GOVERNMENT

Legal systems have evolved in Western democracies as efficient means of resolving disputes by bringing out and evaluating relevant evidence. There are two sides and only two sides in both civil and criminal cases; each side has the task of presenting evidence to support its case and of disputing the evidence presented by the other side. At the conclusion of this process, a judge or jury is responsible for weighing the strength of the evidence presented by both sides and reaching a conclusion in favor of one or the other. Criminal cases occasionally result in hung juries, but this inconclusive outcome is viewed as a failure of the process. This differs fundamentally from science, where there are often more than two competing explanations for a phenomenon and final answers are elusive.

Courts frequently deal with scientific information in reaching their decisions. Chapter 3 discussed two examples in criminal law: the shaky science underpinning the use of dogs in scent identification and the more solid but still imperfect science involved in DNA typing of suspects. Science plays an increasingly important role in civil cases as well, especially in the burgeoning area of toxic torts. These are cases brought by individuals or groups against companies for damages allegedly caused by negligence, often through release of a toxic compound into the environment. These types of cases have been publicized in recent films such as *Erin Brockovich*, a true story about a class action lawsuit against Pacific Gas and Electric Company in California for releasing a cancer-causing form of chromium into water supplies. In some of these cases, the connection between a toxic compound and disease may be indisputable. For example, no one seriously argues against the fact that nicotine causes cancer, and tobacco companies in the United States have finally had to take some financial responsibility for their role in promoting addictive products that cause substantial sickness and death.

In other toxic tort cases, the evidence that specific compounds contained in drugs or released into the environment cause specific diseases is much less clear. This evidence may include geographic clusters of cases of disease that are correlated with elevated concentrations of toxic compounds in drinking water, but such correlational data are often subject to alternative interpretations, as described in Chapter 2. There may also be animal experiments showing that high doses of toxic compounds can produce disease or death in laboratory animals, but these doses are often much higher than humans are exposed to. Another problem with these kinds of cases is that causes of disease in individuals are often multiple and interact in complex ways (see Chapter 6). A certain drug or medical procedure or environmental toxin may increase the risk of a disease, but assessing the significance of this among the panoply of other factors that also influence the disease is a challenging problem that typically requires multiple lines of research. Yet courts regularly make judgements in toxic tort cases based on the incomplete knowledge available at the time of a trial (Jasanoff 1995).

Although the case was settled out of court, the class action suit against the Dow Corning Corporation for selling silicone-gel breast implants illustrates the mismatch that can occur between the demands of the legal system for resolving disputes in a reasonably timely manner and the fact that advances in scientific understanding follow no definite timetable. The suit was based on the hypothesis that implants caused disorders of the immune system, which led to increased risk of arthritis and other diseases of connective tissue. Dow Corning reached a multimillion dollar settlement with the claimants in early 1994. In June 1994, the *New England Journal of Medicine* published a large, retrospective study suggesting that there was no link between silicone-gel breast implants and connective-tissue disease (Angell 1996). The actual health effects of silicone implants are still unresolved (Macklin 1999; Stein 1999).

In passing laws and promulgating rules to implement them, elective bodies and government agencies must also make definitive decisions based on uncertain science. The recent furor in the United States over regulation of arsenic in drinking water is a good example. Arsenic is a toxic metal that can cause cancer, cardiovascular disease, and diabetes. Most Western countries limit the amount of arsenic in drinking water to 10 parts per billion (ppb), but until recently the United States had set the limit at 50 ppb. The U.S. Environmental Protection Agency (EPA), which is responsible for setting rules for safe drinking water, began an intensive review of this limit in 1996 and finally proposed a new standard of 10 ppb in early 2001, just before Bill Clinton was replaced by George W. Bush as president. The EPA review was based in part on an executive order that Clinton had signed early in his presidency to "ensure that the environmental policies [of the EPA] are based on sound science" (Chinni 2001). The primary evidence used to support the new standard was a study of bladder cancer in villagers in southwestern Taiwan who drank well water containing arsenic at levels of greater than 200 ppb, combined with a model to extrapolate from these levels to predict effects of 3 to 20 ppb. This evidence was challenged on two main grounds: that results weren't relevant to the United States because malnutrition among Taiwanese villagers may have exacerbated the damaging effects of arsenic and that the extrapolation from a relatively high level of 200 ppb of arsenic in drinking water to predict risks of arsenic at much lower levels was unjustified.

Early in its term, the Bush administration withdrew the new standard of 10 ppb for arsenic in the United States that had been proposed by the EPA just before Bill Clinton left office. The ostensible reason for this was the claim by Bush and the new administrator of the EPA, Christie Todd Whitman, that scientific controversy about an appropriate allowable level for arsenic in drinking water was still too great to justify a change from 50 to 10 ppb. On 20 March 2001, for example, Bush stated that they would wait to change the limit until "we can make a decision based on sound science" (Chinni 2001). Does this phrase sound familiar?

In fact, the science underlying the original decision by the EPA to lower allowable arsenic levels in drinking water to 10 ppb was quite sound, even though it wasn't absolutely conclusive. The study of Taiwanese villagers, like any research project, rested on certain assumptions, and the conclusions were conditional on those assumptions. Extrapolation of those results to lower levels of arsenic also depended on assumptions, in this case about dose-response relationships of various diseases to toxic metals that have been well established in laboratory experiments with animal model systems. As a regulatory agency, the EPA had to make a specific decision about permissible amounts of arsenic in drinking water within the constraints of these assumptions. In other words, their task was not the intellectual exercise of figuring out precisely how little arsenic it takes to damage human health but the practical problem of setting a reasonable limit for arsenic in drinking water. This decision and all similar decisions are not based solely on science but also involve analyses of costs of new regulations in relation to potential health risks and value judgments about acceptable levels of risk. The Bush administration didn't really rescind the new rule for arsenic of 10 ppb because of concerns about sound science but because the administration weighed the costs to water companies and other businesses of removing excess arsenic more heavily than the previous administration had.

The denouement of this story involved a second review of the evidence on risks of arsenic by the National Academy of Sciences (NAS), which had done an initial review in 1999 that stimulated the EPA to lower the allowable level in January 2001. The NAS summarized four new studies in September 2001 that suggested that risks of consistent use of drinking water with 10 ppb of arsenic were actually two to four times as great (1.3 to 3.7 additional cases of lung and bladder cancer per 1,000 people) as the EPA had estimated in setting the limit at 10 ppb (Kaiser 2001). This (no doubt combined with calculation of political costs and benefits) persuaded the Bush administration to reinstate the new limit of 10 ppb that it had rescinded 6 months earlier.

Political and regulatory decisions have to be based on the scientific knowledge available at specific times, but scientific understanding of many issues continuously evolves. Even a decision to make no decision is still a decision because it means that a current regulation remains in force. For example, *not* reducing the allowable amount of arsenic in drinking water from 50 to 10 ppb implies acceptance of the risks of levels lower than 50 ppb. My fundamental point is that decisions about public policy, like personal decisions about health and nutrition, should be based on a solid understanding of the scientific issues at the time the decision must be made together with careful consideration of other relevant factors, ranging from ethics to economics. I believe that we would be better off if the role of these other factors vis-à-vis scientific factors were more explicit in the decision-making process. Instead, science is often misused or misrepresented in justifying policy decisions.

Another use of the phrase "sound science" by the Bush administration illustrates this point. After many years of research, as well as political wrangling, the federal government finally decided in spring 2002 to proceed with development of Yucca Mountain in southern Nevada as a national repository for nuclear waste that would remain radioactive for thousands of years. Unlike the case of regulating arsenic in drinking water, in which the administration argued that we should *wait* for sound science before changing allowable levels, they claimed that *existing* sound science was the basis for their nuclear waste decision. Yet the scientific uncertainty about the ability of radioactive waste to remain safe for long periods of time at Yucca Mountain is at least as great as uncertainty about the health effects of very low levels of arsenic in drinking water. Although the underground storage site is 300 meters above the water table and very dry now, the area has been wetter in the past and could be wetter in the future, so radioactive material could leach into the water supply of surrounding areas. In addition, the likelihood of volcanic or other geologic activity that could disrupt the storage site has not been fully assessed (Ewing and Macfarlane 2002).

Using phrases like "sound science" to justify political decisions is really a rhetorical device, not a meaningful description of the state of science underlying the decision. This type of political rhetoric is certainly not limited to members of the administration of George W. Bush in the United States but is common among politicians of all political parties in many countries. One unfortunate consequence of this language, however, is that it contributes to public misunderstanding of what science is and how it works. As citizens, all of us need to evaluate as best we can the soundness of scientific information underlying decisions that our elected representatives make about matters of public policy. We need to recognize the strengths and limitations of scientific knowledge relating to these decisions, and we need to remember that the decisions depend on more than science. Most important, they depend on our moral and political philosophy. We would be better served by many politicians and even some scientists if they were more honest about the role of science in relationship to other factors in deciding matters of great public importance.

SCIENCE, ART, AND RELIGION AS WAYS OF KNOWING ABOUT THE WORLD

Science is only one of the methods we have of trying to understand the incredible world in which we live. Two major alternative methods are art and religion. How do these ways of knowing differ? Here is my very personal view of the similarities and differences between science and art and between science and religion.

In comparing science to art, I use the broadest possible definition of art, to include poetry, literature, music, theater, dance, and the humanities in general, as well as the visual arts. On the surface, you may imagine that the approaches of the artist and scientist couldn't possibly be more different. In films, for example, the scientist is often portrayed as logical, rational, and methodical, whereas the artist is portrayed as intuitive, emotional, and inspired. But I believe that many of the apparent differences between science and art are superficial and that there is a fundamental unity in these two approaches to understanding nature. This unity is in the creative spark that is the foundation of both science and art. Scientists and artists share a passion for seeing the world in new ways, for making novel connections between familiar things. Their work involves blood, sweat, and tears, motivated in both cases by apparently irrational faith that they may be onto something really important. The creative process is manifested in very diverse ways. Artistic creativity may often be more intuitive, scientific creativity more systematic. But there is broad overlap in the creative elements of doing science and art, despite big differences in their products.

The most important difference may lie in how the quality of work in these two fields is judged. In science, one criterion of good work is that it stands up to repeated attempts at falsification. Scientific culture entails a competition of ideas, and the ones that are most successful at predicting new observations win out. In this sense, there really is progress in the scientific understanding of the world.[2] Evaluation of good art seems more subjective, and many different artistic expressions of how the world works may coexist. However, another characteristic of the best science is shared with the best art. In both cases, an excellent piece of work can be recognized by the fact that it leads to many new questions and opens up new lines of inquiry. In short, I think the similarities between science and art are more important than the differences. We should cherish and protect both of these quintessentially human passions.

I'm less sanguine about the relationship between science and religion than many scientists. For example, Stephen Jay Gould (1997) has argued that science and religion are "nonoverlapping magisteria," meaning that they are not inherently conflicting because their realms of interest are fundamentally different. Science addresses answerable questions about how the world works; religion addresses questions about ethical behavior—how the world *should* work. However, I'm more interested in how the scientific method of answering questions compares to the religious method, and here I think there is a deep difference between science and the fundamentalist movements that seem so prominent in the world's major religions today, including Christianity, Islam, and Judaism. This difference is that fundamentalism is inherently authoritarian, which is diametrically opposite to the basic workings of science. A favorite maxim of science is "study nature, not books"; in other words, judge evidence relating to an hypothesis based on your own observations and analysis, not what someone tells you. The maxim could be reversed for the fundamentalist movements in the major religious traditions. The Bible, the Koran, or another text contains the Truth. Ordinary people aren't supposed to interpret the text of these religious documents but rather to trust and accept the interpretations of a specially trained priesthood.

I'm obviously skeptical of this authoritarianism, but I must admit that putting faith in authorities on various subjects has an important part in daily life. For example, it usually makes sense to trust your surgeon to operate on you to solve a medical problem, provided you understand all of the treatment options and have reason to believe that your surgeon is competent. Likewise, I trust my auto mechanic to take care of my car properly. This is a matter of living efficiently as much as anything else. Although I couldn't perform surgery on myself, I presumably could learn how to do all the necessary repairs on my car rather than turning it over to an authoritative mechanic. But if I did so, I would have less time for other things that I enjoy doing and find worthwhile.

However, the issues that are *most* important in life are just the ones that we should be *least* willing to turn over to authority figures. These issues include the meaning of life and the difference between right and wrong, traditionally the province of religion. They also include scientific questions that underlie personal, political, and ethical decisions. For example, even though we need to rely on physicians, with their specialized training, for medical care, we have ultimate responsibility for our own health. This doesn't mean ignoring the good advice of physicians or adopting alternative medical treatments just because of a general skepticism of traditional Western medicine. It means taking advantage of the wonderful skills that physicians can provide as teachers and healers but becoming the authorities about our own health. One of my major goals in this book was to give you some of the reasoning tools to do this.

In Chapter 9 I mentioned a famous article by Garrett Hardin (1968) called "The Tragedy of the Commons." Another article by Hardin is less well known but encapsulates for me the fundamental difference between science and religion. In this article he wrote:

> The distinction between science and nonscience is not one of fact but of method. Scientists welcome vulnerability. Others may reject it; if they do, what they produce is not science, and we should say so. . . . It is a paradox of human existence that intellectual approaches claiming the greatest certainty have produced fewer practical benefits and less secure understanding than has science, which freely admits the inescapable uncertainty of its conclusions. (1976:465, 483)

CURIOSITY AND SKEPTICISM

Scientists are a diverse group of people, at least as varied in their personalities as members of any other profession. Some are aggressive; others are docile. There are loners and extroverts. Scientists can be as sensitive, or insensitive, as anyone else. They can be selfish, hoarding their data in fear of being scooped, or wonderfully cooperative. Some are arrogant bullies of their students; others are talented mentors whose influence is wide and deep through their training of the next generation of scientists. However, they have two traits that seem contradictory but characterize almost all successful scientists: curiosity and skepticism. I believe that developing a workable balance between enthusiasm and healthy skepticism about new ideas is one of the primary keys to scientific productivity.

Curiosity is a human quality that seems especially powerful at about age 3, when the pace of learning about language and locomotion, self and society, is so great. For many, curiosity may gradually decline thereafter, but scientists (and artists) are usually people whose curiosity continues to grow into and during adulthood. The creative spark discussed earlier in this chapter and the discoveries described in previous chapters wouldn't be possible without a deep curiosity about how the world works.

Skepticism is equally important for good science, as the means of separating productive ideas from dead ends. It develops later than curiosity and may overwhelm this childlike enthusiasm for newness. A significant element of graduate education in science is teaching students to be critical, and many graduate students adopt this style with gusto. The risk is that they miss the nuggets of beauty and truth in the new research they learn about because of their newfound skill in challenging the assumptions, methods, conclusions, and interpretations of everything they read. The students who go on to become productive scientists are the ones who are able to manage effectively the tension between their curiosity and excitement about all new ideas and their skepticism, which is used to evaluate those ideas and separate the wheat from the chaff.

I hope this book has whetted your curiosity about a variety of topics in biology and medicine. There is much more to learn about these topics; all of the chapters of this book are really just progress reports, and I expect that you will learn about surprising new developments in many of the areas discussed here in the years to come. I hope that I have also given you some tools for reading science news more critically. Perhaps you will be more likely to question the authoritative statements of experts. You may ask yourself, What are the hidden assumptions? What other hypotheses could explain the results? Are there flaws in the experiments or statistical analyses? Have the experts overinterpreted their data? If you can find the right balance between curiosity and skepticism, I believe that your personal decisions involving science and your contributions to social decisions involving science will be more cogent and more satisfying.

RECOMMENDED RESOURCES

Loehle, C. 1996. *Thinking strategically: Power tools for personal and professional advancement.* Cambridge University Press, New York. This is a handbook for developing creativity with applications in science and other fields.

National Institute for Medical Research. 2002. Mill Hill Essays. http://www .nimr.mrc.ac.uk/millhillessays/ (accessed November 30, 2002). Brief essays on a variety of scientific topics for nontechnical readers.

van Gelder, T. 2002. Critical thinking on the web. http:/www.philosophy .unimelb.edu.au/reason/critical/ (accessed November 30, 2002). Many resources for helping you to develop better skills at critical thinking.

Appendix 1

Using Data for Twins to Estimate Genetic and Environmental Contributions to the Risk of Cancer

For monozygotic twins, the correlation in phenotypic values of a trait (r_{MZ}) can be expressed in terms of the genetic variance, the shared environmental variance, and the nonshared environmental variance as

$$r_{MZ} = \frac{V_G + V_{C(MZ)}}{V_G + V_{C(MZ)} + V_{E(MZ)}}.$$

The denominator of the righthand side of this equation represents the total phenotypic variation among all monozygotic twins in the sample. The numerator represents the portion of that variation that is due to differences between different *pairs* of twins because of differences in genetics and shared environments (Adam and Alex grew up on a farm in Iowa; Jason and Jonathan grew up in a tenement house in Chicago). The extra term in the denominator, $V_{E(MZ)}$, is the variation due to nonshared environmental factors, that is, differences *within* pairs of twins. The smaller the value of $V_{E(MZ)}$ in comparison to $V_{G(MZ)}$ and $V_{C(MZ)}$, the more similarity there will be between individuals and their monozygotic twins, producing a larger correlation coefficient for the phenotypic values of the pairs.

For dizygotic twins, the correlation in phenotypic values is

$$r_{DZ} = \frac{(1/2)V_G + V_{C(DZ)}}{V_G + V_{C(DZ)} + V_{E(DZ)}}.$$

The main difference between this equation and the one for monozygotic twins is that the numerator contains the term $(1/2)V_G$ in place of V_G. The reason is that dizygotic twins share only half their genes on average, so the genetic variation between pairs of twins is half of the total genetic variation among all twins, the remainder being variation within pairs. The other difference is that the shared and nonshared environmental variances may not be the same for monozygotic and dizygotic twins; that is, $V_{C(MZ)}$ does not necessarily equal $V_{C(DZ)}$, and $V_{E(MZ)}$ does not necessarily equal $V_{E(DZ)}$. For example, dizygotic twins have separate placentas during fetal development, but about two-thirds of monozygotic twins share a placenta. This suggests that environmental variation related to events during gestation might be greater within pairs of dizygotic twins, so V_E would be relatively greater for dizygotic twins and V_C would be relatively greater for monozygotic twins. However, suppose we assume that shared environmental variances are the same for the two types of twins, as are nonshared environmental variances; that is, $V_{C(MZ)} = V_{C(DZ)} = V_C$ and $V_{E(MZ)} = V_{E(DZ)} = V_E$. This assumption makes it possible to solve the two equations in order to express the components of variance in terms of correlations between twins. The solutions are

$$V_G/V_P = 2(r_{MZ} - r_{DZ});$$
$$V_C/V_P = 2r_{DZ} - r_{MZ};$$
$$V_E/V_P = 1 - r_{MZ}.$$

The beauty of these equations is that they provide a way to estimate the relative contributions of genetic and environmental factors to variation in a trait, such as height or susceptibility to cancer, from two things we can measure—correlations between monozygotic and dizygotic twins in expression of the trait. However, we've made enough assumptions on the way to this point that you may wonder if it has been simply an academic exercise. One justification of the assumption that environmental variances are the same in monozygotic and dizygotic twins is that any differences are probably relatively small, so the equations derived above are approximately correct. In general, every story in science rests on assumptions, although news accounts and even introductory textbooks don't always tell readers what they are. Without assumptions there can be no progress in understanding. Thus, making assumptions is essential, but so is their continuous evaluation.

Substituting $r_{MZ} = 0.366$ and $r_{DZ} = 0.255$ into the equations above, we can estimate that 28% of the variation in breast cancer risk among female Scandinavian twins was due to genetic factors, 8% was due to environmental factors shared by members of a pair, and 63% was due to environmental factors unique to each individual. Lichtenstein's group (2000) used a slightly more complex analysis, which accounted for differences in the average ages of twins studied in Sweden, Denmark, and Finland, but came up with similar estimates: 27% for genetic factors, 6% for shared environmental factors, and 67% for nonshared environmental factors.

Appendix 2

Precision and the Power of Statistical Tests

It may seem like numerical hocus-pocus to assign greater weight to more precise estimates from individual studies than to less precise estimates from other studies in doing a meta-analysis. In fact, however, its justification is rooted in one of the most fundamental concepts in statistics. In describing the experimental study by van Dusseldorp and her colleagues, I introduced the idea of a null hypothesis. This is a widely used tool in statistics, although it seems to be a convoluted approach because the null hypothesis is the opposite of the biological hypothesis of interest. For the Dutch researchers, the biological hypothesis was that caffeine causes an increase in blood pressure. They tested this hypothesis *experimentally* by comparing blood pressure of people drinking caffeinated and decaffeinated coffee. Their *statistical* test of this comparison involved thinking about what might happen if caffeine did not affect blood pressure. If this were the case, could chance produce a difference between the caffeinated and decaffeinated conditions as large as was actually observed? In other words, the biological hypothesis that caffeine causes an increase in blood pressure led to a statistical null hypothesis of no difference in blood pressure from regular and decaffeinated coffee. Under the null hypothesis, the probability of getting as large a difference as actually observed was much less than 5%. Therefore, they rejected the null hypothesis. The reason for this roundabout approach is that the biological hypothesis can't be directly tested statistically without specifying how large an effect of caffeine on blood pressure we would expect. To construct a statistical test, we need a quantitative hypothesis. Since we don't have a detailed mechanistic model to explain how daily consumption of a certain amount of caffeine would lead to a specific increment in blood pressure, we have to be content with the more general biological hypothesis that caffeine causes *some* increase in blood pressure, without specifying its size. The opposite, that caffeine causes no increase in blood pressure, is quantitative because it implies that the average pressure when consuming regular coffee equals the average pressure when consuming decaffeinated coffee. Therefore, we can devise a statistical test of this null hypothesis, as outlined here.

The Dutch researchers rejected the null hypothesis for systolic blood pressure because the average difference for subjects drinking regular and decaffeinated coffee was 1.5 units and the probability of getting this large a difference under the null hypothesis—no real difference in the population from which the subjects were drawn—was about 0.2%. In other words, there was a very small probability that the null hypothesis was true. They rejected the null hypothesis for diastolic pressure using similar logic, but the probability of a mistake was about 1.7% in this case. Researchers often use 5% as an acceptable probability of this type of error; that is, if the statistical test suggests that the probability under the null hypothesis of getting results as extreme or more extreme than those actually obtained is less than 5%, one should reject the null hypothesis; otherwise, one should accept it. This is an arbitrary but fairly low cutoff (researchers sometimes use 1%) and reflects the conservative philosophy that we should credit chance variation as the explanation for most patterns unless there is strong evidence to the contrary. Suppose researchers generally used a cutoff of 25% instead of 5%; this would suggest that about one-quarter of the results of various kinds of experiments reported in the literature did not represent real biological effects but rather just chance differences between treatment and control groups. As a result, there would be a lot of misleading information floating around to confuse other researchers, as well as the general public. At a cutoff of 5% for accepting or rejecting the null hypothesis, it may be somewhat easier to separate the wheat from the chaff of research results. Of course, this all presupposes that experiments are flawless, so results are unbiased. In reality, plenty of problems with the design and execution of experiments force scientists to evaluate the multitude of research results that are published every year.

The most important general point in the last paragraph seems simple but has profound implications. Based on statistical tests, researchers make decisions to reject or accept hypotheses, but these decisions may be wrong. The Dutch researchers rejected the null hypothesis of no effect of caffeine on blood pressure even though there was a small probability that it was true and the differences they found were due solely to chance. This possibility of error is an unavoidable cost of the necessity to test hypotheses like these statistically. However, there is a second type of error that can occur in statistical tests—accepting a false null hypothesis. Table A2.1 summarizes the situation. The null hypothesis may be true or false. After doing an appropriate statistical test on the results of an experiment, we may accept or reject the null hypothesis. If we accept a false null hypothesis or reject a true null hypothesis, our interpretation of the results will be mistaken. These kinds of errors

Table A2.1. Evaluation of hypotheses by statistical tests.

		Status of Null Hypothesis in the Real World	
		True	False
Decision about Null Hypothesis	Accept		Type II Error
as a Result of Statistical Test	Reject	Type I Error	

are eventually corrected by replication of the experiment or by other kinds of evidence brought to bear on the hypothesis, but errors resulting from statistical tests, as outlined in Table A2.1, cause confusion and may delay progress in understanding a phenomenon. It's essential to realize, however, that these are not errors in how the statistical tests were done but rather inherent limitations of this approach to testing hypotheses. Statistics, in particular, and science, in general, can't provide absolute certainty.

A Type II error, in which a false null hypothesis is accepted (Table A2.1), means that there really is an effect of caffeine on blood pressure, for example, but conditions of the experiment were such that the researchers were unable to detect it. In statistical jargon, the experiment didn't have much power to detect the effect. The term "power" has a precise definition in statistics as the probability of *not* making a Type II error; that is, power equals 1 minus the probability of accepting the null hypothesis if it is false. This technical meaning flows naturally from everyday use of the term. A powerful test of an alternative to the null hypothesis—for example, that caffeine consumption will raise systolic blood pressure by a specific amount—is a test that has a high probability of demonstrating this effect if it really exists; that is, of rejecting the false null hypothesis. Biomedical scientists are generally interested in exploring real biological phenomena, not simply chance variation in the natural world, so they want to have powerful tests of their hypotheses.

Before conducting their experiment to compare blood pressure while subjects were drinking regular and decaffeinated coffee, van Dusseldorp and her Dutch colleagues estimated the number of subjects they would need to have 85% power to detect a difference of 2 units in blood pressure between the two conditions.[1] This number was 46 subjects, almost exactly the same as the number (45) actually used. Many of the other studies summarized in the meta-analysis of caffeine and blood pressure by Jee's group (1999) had much lower power than this. In general, the precision with which the size of an effect can be estimated is inversely related to the power of an experiment to document the effect. Wide confidence intervals for several of the studies illustrated in Figure 8.5 mean that these studies had low precision and therefore low power to detect an effect of caffeine on blood pressure. The two major factors that influence the power of an experiment are the sample size and the amount of variation among individuals caused by factors other than the experimental treatment. When sample size is low and variation is high, power and precision are both low.

These considerations of power and precision are why meta-analysts give different weights to various studies in estimating an overall average value for the biological effect of some factor. Studies with low precision and wide confidence intervals are given less weight in estimating the overall size of an effect. Because of their design, these studies have relatively high probabilities of producing a Type II error, acceptance of a false null hypothesis (and thus rejection of the true, alternative hypothesis). Since the chance of error is larger in these studies than in more powerful studies, it makes sense to attach less importance to the weaker studies in assessing the overall consensus about the phenomenon of interest.

Notes

Chapter 1

1. I'll use endnotes like this to elaborate on some points, to provide tidbits of information that may be interesting but would interrupt the flow of the main text, and for other miscellaneous purposes. I deliberately use the phrase "haphazardly selected" because I had the idea to do this while a month's worth of newspapers was ready to be picked up by the recycling truck. I hurried out and pulled off the top bag, picked out the newspapers that had been stuffed in it in no particular order, and noted the information about science stories in the 13 newspapers I had time to look at before the truck arrived.

2. Lori Murray, manager of Corporate Communications for Applied Biosystems, provided the costs of this project in a phone conversation on 28 March 2003. Applied Biosystems and Celera Genomics are divisions of the Applera Corporation, and Applied Biosystems produced the DNA-sequencing machines used by Celera Genomics in its work on the human genome.

3. High technology was important for parts of this story, such as dating the fossils and the molecular work that first suggested that whales and their relatives were more closely related to hippos than to other even-toed ungulates. But the most convincing evidence for the evolutionary transformation was the sequence of the fossils themselves. Even the close relationship between whales and hippos has been supported not only by molecular data but also by new fossil discoveries, which allowed detailed comparison of the anklebones of whales, hippos, and other ungulates.

Chapter 2

1. It's important to note that two versions of the initial picture were shown simultaneously, one with some objects missing. Subjects were asked to compare the two

versions and identify the missing objects in the version shown on the right side of the computer screen. Therefore, they had an opportunity to study the pictures, which enhanced their ability to remember objects in later tests.

2. Since Perrig and his colleagues (1997) didn't provide individual data for their subjects, I created simulated data for 442 individuals by assuming that plasma levels of β-carotene followed a normal distribution, with a mean of 0.72 micromoles per liter and a standard deviation of 0.48; that scores on the test of recognition memory followed a normal distribution, with a mean of 1.33 and a standard deviation of 0.76; and that these two variables were correlated at a level of 0.22 (all of these summary statistics were reported by Perrig's group). The normal distribution is a symmetrical, bell-shaped curve that shows a range of values for a variable, such as plasma concentration of β-carotene, on the horizontal axis and the frequency with which those values occur in a population on the vertical axis. Many variables are normally distributed, and it's a common strategy to assume normal distributions for standard statistical analyses. In fact, although Perrig and his colleagues didn't explicitly state that their data were normally distributed, some of their analyses rest on this assumption. The average, or mean, of a normal distribution occurs at the peak of the curve. The standard deviation is a measure of the variability in the data, or spread of the curve. In a normal distribution, about 68% of the values fall within 1 standard deviation of the mean and 95% fall within 2 standard deviations of the mean. See Figure 8.3 for an illustration of data that fit a normal distribution.

3. The 5% cutoff for statistical significance is a standard rule of thumb (some would say dogma) that thousands of students have learned in beginning statistics classes. The rationale for this rule is that, if the probability of observing a pattern purely by chance is greater than 5%, we should attribute the pattern to chance or, in more formal terms, accept the null hypothesis that the pattern is due to chance. If the probability of observing a pattern purely by chance is less than 5%, we may provisionally accept the hypothesis that there is a real relationship between the variables (see Appendix 2 for more details).

4. The probability that individual A does not get 10 heads is 0.999, the probability that B does not get 10 heads is 0.999, and so on. The probability that A and B and C and so on all do not get 10 heads is $0.999 \times 0.999 \times 0.999 \times \ldots = 0.999^{100} = 0.905$. Therefore, the probability that at least one individual gets 10 heads is $1 - 0.905 = 0.095$.

5. "Confident" here means that the probabilities of getting these correlations by chance alone are less than 5% in the context of the whole analysis (see Appendix 2). A dramatic example of the pitfalls of interpreting multiple statistical tests of the same set of data occurred in the spring of 2003 when VaxGen Incorporated reported the first tests of an AIDS vaccine they had developed. In their overall analysis of results for 5,000 subjects, they found no evidence that the vaccine prevented infection with HIV, the virus that causes AIDS. But VaxGen researchers reported a significant benefit in blacks when they analyzed data separately for nine ethnic subgroups. However, they apparently failed to apply a Bonferroni correction to these analyses. Had they done so, the probability that the benefit seen in blacks was due to chance would have increased from less than 2% to between 9% and 18%, too high to have much confidence that there was a real benefit of the vaccine for that group (Cohen 2003).

6. Cognitive impairment caused by a succession of small strokes.

7. A meta-analysis is a formal, statistical procedure for combining the results of multiple studies of the same question to arrive at an objective consensus about the answer (see Chapter 8).

8. The subjects in the initial trial were mostly white females with an average age of 36 and no other diseases. The FDA Advisory Committee wanted to see results for a broader range of subjects and more information on side effects before recommending approval. The committee also noted that pleconaril had to be started within the first 24 hours of onset of a cold to be effective and wondered about the practicality of this treatment (Food and Drug Administration 2002; Senior 2002).

Chapter 3

1. In 1993 the U.S. Supreme Court decided that the Frye Rule, which was promulgated in 1923, was superseded by the Federal Rules of Evidence, which became the legal basis for admissibility of evidence in federal cases in 1975. These rules are used in many state courts, as well as by federal judges. In its 1993 decision, the Supreme Court also specified four criteria for evaluating scientific evidence in court: (1) the methods underlying the evidence should be testable, (2) published or reviewed by scientific peers, and (3) generally accepted by the relevant community of scientists, and (4) the likelihood of error in the methods should be able to be estimated (Annas 1994). To my knowledge, this Supreme Court decision, *Daubert v. Merrell Dow Pharmaceuticals*, has not been applied to the use of dogs for scent identification in police lineups.

2. See Chapter 6 for a discussion of more precise terminology for classifying types of twins.

3. In a recent study, Wells and Hepper (2000) tested the ability of *humans* to distinguish between the scents of different *dogs*. They found that dog owners could pick out a blanket scented by their own dog when it was paired with a blanket scented by an unfamiliar dog, although they expressed no consistent preference between the two.

4. Schoon published her dissertation research in articles in *Applied Animal Behaviour Science* (1996), *Behaviour* (1997), and the *Journal of Forensic Science* (1998).

5. A horse named Clever Hans is a famous example of the ability of animals to attend to very inconspicuous movements of humans. Clever Hans, who lived in Germany in the early 1900s, purportedly could solve mathematical and other problems and indicate correct answers by tapping his hoof. Clever Hans was very popular with the public, especially after a group of scientists saw him demonstrate his talents and declared that there was no evidence of fraud on the part of his owner. This inspired an experimental psychologist named Oskar Pfungst to test Clever Hans in various ways, including the first double-blind experiments in psychology. If the owner didn't know the answer to a question or if a curtain was placed between the owner and Clever Hans, the horse never got the right answer. After long and careful observation, Pfungst finally discovered that the horse noticed barely perceptible and unconscious head movements or postural changes of the owner when the horse had reached the correct number of taps, and he stopped tapping at that point (J. L. Gould 1982).

6. Here is a similar problem to test your understanding of these calculations. Suppose there are two taxi companies in Reno, Blue Cab and Green Cab. Eighty-five percent of the cabs in town are blue while 15% are green. A cab was responsible for a hit-and-run accident at night; an eyewitness saw the accident and reported that the cab was green. The police tested the ability of the witness to determine the color of cabs at night and found that the witness was accurate 80% of the time. What is the probability that the cab involved in the accident was actually green?

7. Substitute one guilty suspect, one innocent suspect, and two total suspects in the bottom row of Table 3.4; then recalculate the top row of this table by multiplying

0.36 by 1 and 0.05 by 1 and adding these two values to get 0.41. Therefore, the probability that the suspect identified by the dog is actually guilty is $0.36/0.41 = 0.88$.

8. Why are these probabilities of a false positive result so low? Laboratories that test DNA typically analyze several different regions of the DNA extracted from a sample of blood or other tissue. For each of these regions, several different sequences of DNA typically exist in a population. Two unrelated individuals might have the same sequence at some of these regions, but the probability that they will have the same sequence at all regions tested is very low because it is the product of the probabilities for each individual region. For example, if the chance that two people have the same DNA sequence at any one tested region is 0.2 and 10 regions are tested, the chance that they have the same sequence at all 10 regions is 0.2^{10}, which equals approximately one in 10 million.

Chapter 4

1. Researchers obviously didn't design experimental tests of the teratogenicity of retinoic acid with human subjects. However, an acne medicine called Accutane, which contains a derivative of retinoic acid as an active ingredient, was used by some pregnant women despite warnings to the contrary. About half of one set of women who were exposed to Accutane during pregnancy either miscarried or had babies with various birth defects.

2. By contrast, imagine a more general hypothesis that some unspecified chemical in the water causes deformities. Perhaps this could be tested by finding a set of ponds where a high percentage of frogs are deformed and another set where most frogs are normal, collecting water from these ponds, and comparing the chemical constituents of these samples. But natural bodies of water differ chemically in hundreds if not thousands of ways, so it would be very difficult to pinpoint the precise chemical that might be responsible for deformities. The problem is compounded by the fact that chemical reactions take place continuously in water so the hypothetical compound that was present in the right concentration to cause abnormal limb development at the time of metamorphosis from tadpoles to frogs might have been converted into something else by the time water samples were taken. This approach is searching for a needle in a haystack.

3. There are a great variety of methods for estimating the sizes of natural populations. Some of these are fairly straightforward, such as counting the number of males displaying at a pond during the breeding season and using this as an index of abundance; others involve sophisticated mathematical analysis of resightings or recaptures of marked animals. Regardless of whether estimates come from a simple or complicated method, the key requirement is that methods for each population must be consistent over time. It's not necessary for the validity of the analysis by Houlahan et al. (2000) that all researchers used the same method, only that they didn't change how they estimated the size of a particular population in midstream.

4. Conceivably the hypothesis that the pathogenic fungus was transferred from hatchery-reared fish to frogs and toads could be tested by selecting several pristine lakes without such fish, introducing fish infected with the fungus to a random subset of these lakes, and introducing uninfected fish to the remaining lakes as controls. Then the egg survival and population dynamics of amphibians that are breeding in these lakes could be followed for several years. But the ethics of doing this experiment are questionable, considering the threatened status of some of the amphibian species.

Chapter 5

1. Fledglings are baby birds that have left the nest and are able to fly but may still be dependent on their parents.

2. Since the area of an average cache is 2 square centimeters and caches are roughly circular in shape, the radius of a cache is $\sqrt{2/\pi}$. This is also the radius of a typical probe. If we assume that a bird finds a cache if the edge of a probe hits the edge of the cache, the area a bird has to hit to discover a cache includes the area of the cache itself plus a ring of radius $\sqrt{2/\pi}$ outside the cache. Therefore, the radius of the total area that will result in discovery of a cache equals $2\sqrt{2/\pi}$, and this area equals 8 square centimeters. With 379 caches, the area that a bird has to hit with a probe to discover any cache equals $379 \times 8 = 3032$ square centimeters. The total area of the arena is 750,000 square centimeters, so the chance of discovering a cache by random search equals $3032/750,000 = 0.004$, or 0.4%.

3. Why did two of the nutcrackers used by Vander Wall (1982) not store seeds in the arena? The most likely possibility is that they did not adapt as well to captivity as the two birds that did cache. Another possibility, however, is that different birds have different strategies under natural conditions: some might store their own seeds; others might make a living stealing caches made by the first group. Vander Wall argued that this second strategy is unlikely to be successful because the ability to locate caches by using directed random search and microtopographic cues would degrade over time as caches are harvested and cache density decreases and as microtopographic cues are erased by wind, rain, and other weathering processes. Therefore, birds that depended on a pilferage strategy would have the most difficult time finding food during the winter and subsequent breeding season, when their needs might be greatest. Testing these alternatives will require further laboratory experiments with more birds and more detailed studies in the field.

4. The least familiar of these rodents may be kangaroo rats, which are one of my favorite animals. These are so named because they have well-developed hindlimbs and hop around their desert habitats, although they aren't related to kangaroos. They use their delicate forelimbs to sift the sand for seeds, which they carry to burrows and other storage sites in fur-lined cheek pouches with separate openings adjacent to the mouth. They have many additional morphological, physiological, and behavioral adaptations to desert environments.

Chapter 6

1. A fertilized egg is called a zygote. In the early stages of development, this single cell divides to make two cells, the two divide to make four, then eight, and so on. In rare cases in humans, this ball of cells may separate into two balls, each of which develops into an individual. These two individuals have the same genes and thus are commonly called identical twins. They are genetically identical but not identical in an absolute sense because they experience somewhat different environments. Even before birth, for example, they are in different positions in the uterus, which may cause them to develop slightly differently. Therefore, a more accurate designation for identical twins is *monozygotic*—from one zygote. Fraternal twins, on the other hand, arise when two separate eggs are fertilized by different sperm. These two fertilized eggs form two zygotes; thus fraternal twins are *dizygotic*. Their average degree of genetic similarity is no more or less than that of two siblings with the same mother and father born at different times.

2. The term "genetic causation" has at least two meanings: (1) changes in chromosomes or DNA that affect functioning of a cell and (2) inheritance of a form of a gene that influences a trait of an individual, such as his or her susceptibility to a disease. Genetic changes that occur in somatic cells (all cells except sperm in males and eggs in females) may alter how those cells work but are not passed on to offspring. From now on, I'll use "genetic causation" in the second sense.

3. The 24 unique chromosomes are labeled 1–22, X, and Y. Each one can be identified under a microscope by its distinctive size and shape. Each individual has 46 chromosomes in all cells of the body except mature sperm or eggs. A female has two copies of chromosomes 1–22 and two copies of the X chromosome. A male has two copies of chromosomes 1–22 plus one X and one Y.

4. If the mutation rate is 1×10^{-6} to 1×10^{-7} and it takes two mutations to produce a tumor, how can the incidence of sporadic cases be as high as one in 50,000? If we considered only one cell, the probability of two independent mutations would be at most $(1 \times 10^{-6}) \times (1 \times 10^{-6}) = 1 \times 10^{-12}$, or one in 1 trillion. But each human retina has about 2 million cells, which arise by about 18 successive doublings from 10 initial cells in an early embryo. Thus there are many opportunities for these mutations to occur, producing the incidence rate of one in 50,000. I've oversimplified this example somewhat because in fact different types of somatic mutations can occur in the retina to produce retinoblastoma, and the mutation rates are not necessarily the same for the first and second mutations.

5. For this child, only one spontaneous mutation is required in an eye to produce retinoblastoma in that eye. As described in the preceding note, there are many opportunities for this mutation to occur because of the large number of cell divisions that take place during formation of the retina. This results in a very high probability of retinoblastoma in one eye and a substantial probability of the disease in both eyes.

6. Lichtenstein's group (2000) used a slightly more complex method of estimating relative risks, which accounted for the fact that twins of some of the focal individuals had been diagnosed with cancer several years previously whereas twins of other focal individuals had been diagnosed with cancer only recently. Nevertheless, the values they reported for the relative risk of breast cancer were similar to the ones shown here: 5.2 for monozygotic twins and 2.8 for dizygotic twins.

7. The implications of relative risk are often misrepresented by physicians or misinterpreted by patients. As discussed by Dupont and Plummer (1996), the important statistic for a person concerned about the chances of getting a specific disease is *absolute risk*, which is the probability of getting the disease during a certain period of time, like the next 10 years. An individual may have a high relative risk compared to the general population, but if the frequency of the disease in the general population is very low, his or her absolute risk is still low.

8. One type of data that has *not* provided much evidence for environmental causes of cancer is examination of cancer clusters in communities. These are situations in which multiple cases of cancer are reported in a small geographic area. This would seem to be an ideal opportunity for linking a specific environmental factor to a specific type of cancer, by identifying common exposure to a specific factor among afflicted individuals in the community. Despite many requests from the public to investigate cancer clusters (e.g., 1,500 in the United States in 1989), Trichopoulos and his colleagues (1996) reported that only one localized cancer cluster had ever been definitively linked to a specific environmental factor. This was the discovery that multiple cases of a rare form of respiratory cancer in a village in Turkey could be attributed to a mineral similar to asbestos found in the soil. There are many reasons that apparent

cancer clusters perceived by the public and discussed in the press are rarely conclusively tied to specific environmental causes. The most basic reason is that cancer is a common set of diseases and there are many communities in a large country, such as the United States, so there are many opportunities for concentrations to occur in specific communities simply by chance (Robinson 2002).

Chapter 7

1. In Chapter 6, I discussed the contribution to retinoblastoma and breast cancer of genes that specify tumor-suppressor proteins. The p53 gene is probably the most important; it is altered by mutation in at least half of human cancers, resulting in a form of the p53 protein that is inactive.

2. Many, but not all, tissues and organs experience continuous turnover of cells, so the ages of the characteristic cells of these tissues and organs are roughly constant even though the bodies in which they occur keep getting older. One example is red cells in the blood. Red blood cells have an average life span of 120 days in humans; new red cells are continuously formed from undifferentiated stem cells in the bone marrow to replace mature red cells that are continuously lost from the blood.

3. The probability of living from birth to age 1 is 0.99; from age 1 to age 2, 0.99; from age 2 to age 3, 0.99; and so on. Therefore, the probability of living from birth to age 2 is $0.99 \times 0.99 = 0.99^2 = 0.98$. The probability of living from birth to age 50 is 0.99^{50}, and the probability of living from birth to age 100 is 0.99^{100}.

4. Connective tissue takes many different forms in the body. Some of the most familiar are tendons, ligaments, and the walls of blood vessels, but other forms of connective tissue occur throughout the body. For example, fat is composed of connective tissue.

5. Fifty cell divisions, or 50 successive doublings, produce a very large number of cells: 2^{50} equals about 1.13×10^{15} cells, or 1 million billion cells.

6. You began life as a fertilized egg produced in one of your mother's ovaries as one of the trillions of cells in her body that arose from the fertilized egg that would eventually become your mother. That egg in turn was produced in one of your maternal grandmother's ovaries: the chain can be carried backward or forward in time as far as you wish.

7. Wild-type worms are descendants of worms collected in nature and subsequently maintained in the laboratory with no exposure to agents that cause mutations.

8. Although primordial germ cells originate outside the embryo, they are ultimately derived from the fertilized egg and therefore are genetically identical to all the other cells of the embryo. As the fertilized egg begins to divide, it forms a ball of cells called a blastocyst that becomes the developing embryo, as well as a set of membranes that contribute to the placenta. The primordial germ cells are derived from one of these membranes.

9. The probability of surviving for 1 month is $1 - 0.1 = 0.9$. The probability of surviving for 1 year is $0.9 \times 0.9 \times 0.9 \times \ldots = 0.9^{12} = 0.282$, or 282 mice of the 1,000 we started with. The other calculations are similar.

Chapter 8

1. Blood pressure is typically measured at two times during a complete cycle of contraction and relaxation of the heart. Systolic pressure is the pressure of blood in the arteries when the ventricles of the heart are contracting and expelling blood; diastolic pressure is the pressure when the heart muscle is relaxed and the ventricles are

being refilled with blood. Blood pressure is expressed in millimeters of mercury (mm Hg), a standard way of measuring pressure in general (e.g., atmospheric pressure at sea level is 760 mm Hg). For simplicity, I'll often refer to these measurements as "units." For subjects in the Dutch experiment, the average baseline blood pressure was 124/76 (systolic/diastolic). All subjects had baseline values in the normal range, which is less than 140 mm Hg for systolic pressure and less than 90 mm Hg for diastolic pressure.

2. Note the change in wording from Giovannucci's (1999) cautious "associated with a lower risk" to Ekbom's (1999) assertive "are at lower risk." Ekbom's remark about being "forced into a life without" coffee refers to an experiment with two prisoners in Sweden in the eighteenth century. One was given coffee daily; the other was not. Neither prisoner had colorectal cancer when he was released from prison after 20 years. Thus my comments about the impossibility of long-term experiments aren't *quite* justified.

Chapter 9

1. The Celsius temperature scale is used in most countries and in all scientific writing and discussion. For American readers who may not be used to thinking in these units, a temperature change of 1 degree Celsius translates to 1.8 degrees Fahrenheit; that is, to convert a change in °C to °F, multiply by 1.8 (multiply by 2 for a rough approximation). Thus a change of 33°C would be 59°F. If we are converting specific temperatures, we have to account for the fact that 0°C corresponds to 32°F. For example, normal body temperature of most mammals is about 37°C, or $(1.8 \times 37) + 32 = 98.6$°F.

2. Although this position is fundamentally conservative, in the sense of being restrained or cautious, it's the opposite of the position taken by many political conservatives, at least in the United States.

3. This definition of a model as an hypothesis with explicit assumptions about specific and detailed relationships is somewhat limited, although useful for our present purposes. Some models incorporate information or concepts from multiple disciplines, whereas we usually think of hypotheses as being part of just one discipline. Models are often idealized and purposely unrealistic, and they may have more value as a heuristic tool than in making accurate predictions that can be tested. By defining a model as a type of hypothesis for purposes of this discussion, I don't want to imply disagreement with those who define models more broadly.

4. Believe it or not, Martens and his colleagues (1999) named their model MIASMA—*M*odelling framework for the health *I*mpact *AS*sessment of *M*an-induced *A*tmospheric changes.

5. Degree-days are used to represent the heat requirements for biological processes such as plant growth and animal development under natural conditions. For example, if an early-maturing variety of tomato plant is well watered and fertilized, it takes about 1,100 Celsius degree-days for tomatoes to develop and ripen, where the minimum temperature for the process is 10°C. If the average daily temperature during the growing season is 25°C (77°F), you should have ripe tomatoes about 73 days after planting Early Girls or a similar variety because $73 \times (25$°C $- 10$°C$) = 1095$ degree-days.

6. George Macdonald (1961) introduced the concept of the basic reproduction rate in the 1950s, and Robert May (1983) estimated basic reproduction rates for malaria of 16 to 80 in Nigeria, where the disease is common. Why, then, did Martens and his

colleagues (1999) use a relative measure of malaria risk in their model, rather than directly estimating how global climate change would affect the basic reproduction rate? May had used indirect methods to estimate basic reproduction rates for malaria in Nigeria, but these methods only work for areas where a disease already exists, so they can't be used to predict how climate change might influence the spread of malaria to a new area. Transmission potential, used by Martens's group as a relative measure of risk, includes only the terms in the full formula for the basic reproduction rate that are known to depend on temperature. Therefore, transmission potential equals the biting rate of mosquitoes times the probability that an infected mosquito lives long enough for the parasite to complete its incubation period (Figure 9.4). This assumes that the following elements of the basic reproduction rate are independent of temperature: the size of the mosquito population relative to that of the human population in an area, the length of time a person with malaria is infectious to mosquitoes, the probability of successful transfer of parasites from an infected person to a mosquito, and the probability of successful transfer of parasites from an infected mosquito to a person. As discussed in the text, the least plausible of these assumptions is probably the first, but Martens's group believed this assumption to be conservative.

7. I've simplified this example in four major ways: (1) In a mass vaccination program for smallpox, the vaccine would not be offered to people with elevated risks, including very young children; pregnant women; and people with cancer, AIDS, or the skin disease eczema. (2) Since the vaccine is a live virus, it can be transmitted from a vaccinated person to an unvaccinated person through close contact, so people living with those in high-risk categories would also not be vaccinated. (3) These two groups of people make up about 25% of the U.S. population, which means that everyone else would have to be vaccinated to prevent the spread of smallpox. (4) The calculations of vaccination risks made by Kemper and his colleagues (2002) were based on historical data before 1972, when mass vaccination was done in the United States. Risks today might be lower, if safer vaccines could be developed, or higher, if more people have compromised immune systems than before 1972.

Chapter 10

1. One and one-half million is a minimum estimate of the number of species that have been identified and given scientific names. This is probably fewer than 50% of the total number of species that exist on Earth since new ones are being discovered continuously, especially in the tropics. Some workers have estimated that 100 million different species occur on Earth (Wilson 2002).

2. I am taking the traditional, positivist approach of the scientist here. Some would argue vociferously that all scientific knowledge is socially mediated, so it is meaningless to talk of progress in scientific understanding. Ronald Giere (1999) analyzes this debate in *Science without Laws*.

Appendix 2

1. Explaining methods for calculating statistical power is beyond the scope of this book. Van Dusseldorp and her colleagues (1989) didn't give enough details for me to reproduce their calculations; it's probably safe to assume, however, that they used an effect size of 2 units of blood pressure because previous work had suggested that if average blood pressure was reduced by this amount in people in industrialized countries, there would be a substantial reduction in mortality from cardiovascular disease (see

the discussion of the paradox of prevention in Chapter 8). It's also important to note that power is linked to the probability of Type I error (rejection of a true null hypothesis) that guides a study. The Dutch researchers used a probability of Type I error of 5% for their work. If they had used a smaller probability of Type I error, the probability of Type II error would have been greater; that is, they would have had to be satisfied with lower power.

References

Ammon, H. P. T., P. R. Bieck, D. Mandalaz, and E. J. Verspohl. 1983. Adaptation of blood pressure to continuous heavy coffee drinking in young volunteers: A double-blind crossover study. *British Journal of Clinical Pharmacology* 15:701–706.

Andrieu, N., and A. M. Goldstein. 1998. Epidemiologic and genetic approaches in the study of gene-environment interaction: An overview of available methods. *Epidemiologic Reviews* 20:137–147.

Andronova, N. G., and M. E. Schlesinger. 2001. Objective estimation of the probability density function for climate sensitivity. *Journal of Geophysical Research-Atmospheres* 106:22605–22611.

Angell, M. 1996. *Science on trial: The clash of medical evidence and the law in the breast implant case.* Norton, New York.

Ankley, G. T., J. E. Tietge, D. L. DeFoe, K. M. Jensen, G. W. Holcombe, E. J. Durham, and S. A. Diamond. 1998. Effects of ultraviolet light and methoprene on survival and development of *Rana pipiens*. *Environmental Toxicology and Chemistry* 17:2530–2542.

Annas, G. J. 1994. Scientific evidence in the courtroom—the death of the Frye Rule. *New England Journal of Medicine* 330:1018–1021.

Antithrombotic Trialists' Collaboration. 2002. Collaborative meta-analysis of randomised trials of antiplatelet therapy for prevention of death, myocardial infarction, and stroke in high risk patients. *British Medical Journal* 324:71–86.

Armstrong, D. P. 1991. Levels of cause and effect as organizing principles for research in animal behaviour. *Canadian Journal of Zoology* 69:823–829.

Audera, C., R. V. Patulny, B. H. Sander, and R. M. Douglas. 2001. Mega-doses of vitamin C in treatment of the common cold: A randomised controlled trial. *Medical Journal of Australia* 175:359–362.

Austad, S. N. 1993. Retarded senescence in an insular population of Virginia opossums (*Didelphis virginiana*). *Journal of Zoology, London* 229:695–708.

————. 1997. *Why we age: What science is discovering about the body's journey through life.* Wiley, New York.

Bak, A. A. A., and D. E. Grobbee. 1989. Abstinence from coffee leads to a fall in blood pressure. *Journal of Hypertension* 7:S260–S261.

Balda, R. P., and A. C. Kamil. 1989. A comparative study of cache recovery by three corvid species. *Animal Behaviour* 38:486–495.

————. 1992. Long-term spatial memory in Clark's nutcracker, *Nucifraga columbiana. Animal Behaviour* 44:761–769.

Bartlett, R. H., D. W. Roloff, J. R. Custer, J. G. Younger, and R. B. Hirschl. 2000. Extracorporeal life support: The University of Michigan experience. *JAMA* 283: 904–908.

Beckman, K. B., and B. N. Ames. 1998. The free radical theory of aging matures. *Physiological Reviews* 78:547–581.

Begg, C. B. 2001. The search for cancer risk factors: When can we stop looking? *American Journal of Public Health* 91:360–364.

Blaustein, A. R. 1994. Amphibians in a bad light. *Natural History* 103(10):32–39.

Blaustein, A. R., P. D. Hoffman, D. G. Hokit, J. M. Kiesecker, S. C. Walls, and J. B. Hays. 1994a. UV repair and resistance to solar UV-B in amphibian eggs: A link to population declines? *Proceedings of the National Academy of Sciences USA* 91: 1791–1795.

Blaustein, A. R., D. G. Hokit, R. K. O'Hara, and R. A. Holt. 1994b. Pathogenic fungus contributes to amphibian losses in the Pacific Northwest. *Biological Conservation* 67:251–254.

Blaustein, A. R., and P. T. J. Johnson. 2003. The complexity of deformed amphibians. *Frontiers in Ecology and the Environment* 1:87–94.

Blaustein, A. R., and J. M. Kiesecker. 2002. Complexity in conservation: Lessons from the global decline of amphibian populations. *Ecology Letters* 5:597–608.

Blaustein, A. R., J. M. Kiesecker, D. P. Chivers, D. G. Hokit, A. Marco, L. K. Belden, and A. Hatch. 1998. Effects of ultraviolet radiation on amphibians: Field experiments. *American Zoologist* 38:799–812.

Blaustein, A. R., J. M. Kiesecker, D. G. Hokit, and S. C. Walls. 1995. Amphibian declines and UV radiation. *Bioscience* 45:514–515.

Blaustein, A. R., and D. B. Wake. 1995. The puzzle of declining amphibian populations. *Scientific American* 272(4):52–57.

Bouma, M. J., C. Dye, and H. J. Van Der Kaay. 1996. Falciparum malaria and climate change in the Northwest Frontier Province of Pakistan. *American Journal of Tropical Medicine and Hygiene* 55:131–137.

Braun, B. L., J. B. Fowles, L. Solberg, E. Kind, M. Healey, and R. Anderson. 2000. Patient beliefs about the characteristics, causes, and care of the common cold: An update. *Journal of Family Practice* 49:153–156.

Brisbin, I. L., Jr., and S. N. Austad. 1991. Testing the individual odour theory of canine olfaction. *Animal Behaviour* 42:63–69.

Brisbin, I. L., Jr., S. Austad, and S. K. Jacobson. 2000. Canine detectives: The nose knows — or does it? *Science* 290:1093.

Bruton, S., and F. Fong. 2000. *Science content standards for California public schools: Kindergarten through Grade 12.* California Department of Education, Sacramento.

Burr, M. L., J. E. J. Gallacher, B. K. Butland, C. H. Bolton, and L. G. Downs. 1989. Coffee, blood pressure and plasma lipids: A randomized controlled trial. *European Journal of Clinical Nutrition* 43:477–483.

Carter, K. C. 1985. Koch's postulates in relation to the work of Jacob Henle and Edwin Klebs. *Medical History* 29:353–374.

Chalmers, T. C. 1975. Effects of ascorbic acid on the common cold: An evaluation of the evidence. *American Journal of Medicine* 58:532–536.

———. 1996. Dissent to the preceding article by H. Hemilä. *Journal of Clinical Epidemiology* 49:1085.

Chamberlin, T. C. 1890. The method of multiple working hypotheses. *Science* (old series) 15:92–96 (reprinted in *Science* 148:754–759 [1965]).

Chinni, D. 2001. Arsenic flap and 'sound science.' *Christian Science Monitor,* 14 June 2001.

Christie, M. 2000. *The ozone layer: A philosophy of science perspective.* Cambridge University Press, Cambridge.

Clausen, J., D. D. Keck, and W. M. Hiesey. 1948. *Experimental studies on the nature of species. III. Environmental responses of climatic races of Achillea.* Carnegie Institution of Washington Publication No. 581, Washington, D.C.

Cohen, J. 2003. Vaccine results lose significance under scrutiny. *Science* 299:1495.

Colditz, G., W. Dejong, D. Hunter, D. Trichopoulos, and W. Willett. 1996. Harvard report on cancer prevention, volume 1: Causes of human cancer. *Cancer Causes and Control* 7:S3–S59.

Colditz, G. A., W. Dejong, K. Emmons, D. J. Hunter, N. Mueller, and G. Sorensen. 1997. Harvard report on cancer prevention, volume 2: Prevention of human cancer. *Cancer Causes and Control* 8:S1–S50.

Congdon, J. D., R. D. Nagle, O. M. Kinney, and R. C. van Loben Sels. 2001. Hypotheses of aging in a long-lived vertebrate, Blanding's turtle (*Emydoidea blandingii*). *Experimental Gerontology* 36:813–827.

Culotta, E. 1991. Science's 20 greatest hits take their lumps. *Science* 251:1308–1309.

De Angelis, H., and P. Skvarca. 2003. Glacier surge after ice shelf collapse. *Science* 299:1560–1562.

Douglas, R. M., E. B. Chalker, and B. Treacy. 2001. *Vitamin C for preventing and treating the common cold (Cochrane Review).* The Cochrane Library, Issue 4, Update Software, Oxford.

Draper, G. J., B. M. Sanders, P. A. Brownbill, and M. M. Hawkins. 1992. Patterns of risk of hereditary retinoblastoma and applications to genetic counselling. *British Journal of Cancer* 66:211–219.

Dupont, W. D., and W. D. Plummer, Jr. 1996. Understanding the relationship between relative and absolute risk. *Cancer* 77:2193–2199.

Eggertsen, R., A. Andreasson, T. Hedner, B. E. Karlberg, and L. Hansson. 1993. Effect of coffee on ambulatory blood pressure in patients with treated hypertension. *Journal of Internal Medicine* 233:351–355.

Eisenberg, J. F., and D. E. Wilson. 1978. Relative brain size and feeding strategies in the Chiroptera. *Evolution* 32:740–751.

Ekbom, A. 1999. Commentary [Review: Substantial coffee consumption was associated with a lower risk of colorectal cancer in the general population]. *Gut* 44:597.

Enserink, M. 2002. The vanishing promises of hormone replacement. *Science* 297:325–326.

Epstein, P. R. 2000. Is global warming harmful to health? *Scientific American* 283(2): 50–57.

———. 2001. Climate change and emerging infectious diseases. *Microbes and Infection* 3:747–754.

Evans, P., and B. Halliwell. 2001. Micronutrients: Oxidant/antioxidant status. *British Journal of Nutrition* 85:S67–S74.

Ewing, R. C., and A. Macfarlane. 2002. Yucca Mountain. *Science* 296:659–660.

Faddy, M. J., R. G. Gosden, A. Gougeon, S. J. Richardson, and J. F. Nelson. 1992. Accelerated disappearance of ovarian follicles in mid-life: implications for forecasting menopause. *Human Reproduction* 7:1342–1346.

Finch, C. E., and T. B. L. Kirkwood. 2000. *Chance, development, and aging*. Oxford University Press, New York.

Finch, C. E., M. C. Pike, and M. Witten. 1990. Slow mortality rate accelerations during aging in some animals approximate that of humans. *Science* 249:902–905.

Food and Drug Administration, U.S. Department of Health and Human Services. 2002. *Transcript of meeting of the Antiviral Drugs Advisory Committee on 19 March 2002*. http://www.fda.gov/ohrms/dockets/ac/02/transcripts/3847t1.pdf (accessed April 1, 2003).

Food and Nutrition Board, Institute of Medicine, U.S. National Academy of Sciences (Panel on Dietary Antioxidants and Related Compounds, Subcommittees on Upper Reference Levels of Nutrients and Interpretation and Uses of Dietary Reference Intakes and Standing Committee on the Scientific Evaluation of Dietary Reference Intakes). 2000. *Dietary reference intakes for vitamin C, vitamin E, selenium, and carotenoids*. National Academy Press, Washington, D. C.

Gannett, L. 1999. What's in a cause?: The pragmatic dimensions of genetic explanations. *Biology & Philosophy* 14:349–374.

Gaulin, S. J. C., and R. W. FitzGerald. 1986. Sex differences in spatial ability: An evolutionary hypothesis and test. *The American Naturalist* 127:74–88.

Giere, R. N. 1997. *Understanding scientific reasoning*, 4th ed. Harcourt Brace College Publishers, Fort Worth, Tex.

———. 1999. *Science without laws*. University of Chicago Press, Chicago.

Giovannucci, E. 1998. Meta-analysis of coffee consumption and risk of colorectal cancer. *American Journal of Epidemiology* 147:1043–1052.

———. 1999. Review: Substantial coffee consumption was associated with a lower risk of colorectal cancer in the general population. *Gut* 44:597.

Glass, G. V. 1977. Integrating findings: The meta-analysis of research. *Review of Research in Education* 5:351–379.

Gomulkiewicz, R., and N. A. Slade. 1997. Legal standards and the significance of DNA evidence. *Human Biology* 69:675–688.

Gould, J. L. 1982. *Ethology: The mechanisms and evolution of behavior*. Norton, New York.

Gould, S. J. 1997. Nonoverlapping Magisteria. *Natural History* 106(2):16–22, 60–62.

Graudal, N. A., A. M. Galløe, and P. Garred. 1998. Effects of sodium restriction on blood pressure, renin, aldosterone, catecholamines, cholesterols, and triglyceride: A meta-analysis. *JAMA* 279:1383–1391.

Griffin, D. R. 1986. *Listening in the dark: The acoustic orientation of bats and men*. Cornell University Press, Ithaca, N.Y.

Grimes, D. A., and K. F. Schulz. 2002. Uses and abuses of screening tests. *The Lancet* 359:881–884.

Hardin, G. 1968. The tragedy of the commons. *Science* 162:1243–1248.

———. 1976. Vulnerability—the strength of science. *American Biology Teacher* 38: 465, 483.

Harman, D. 1956. Aging: A theory based on free radical and radiation chemistry. *Journal of Gerontology* 11:298–300.

Hartman, T. J., J. A. Tangrea, P. Pietinen, N. Malila, M. Virtanen, P. R. Taylor, and

D. Albanes. 1998. Tea and coffee consumption and risk of colon and rectal cancer in middle-aged Finnish men. *Nutrition and Cancer* 31:41–48.

Hayes, T., K. Haston, M. Tsui, A. Hoang, C. Haeffele, and A. Vonk. 2002. Feminization of male frogs in the wild. *Nature* 419:895–896.

Hayflick, L., and P. S. Moorhead. 1961. The serial cultivation of human diploid cell strains. *Experimental Cell Research* 25:585–621.

Hazen, R. M., and J. Trefil. 1991. *Science matters: Achieving scientific literacy*. Doubleday, New York.

Hemilä, H. 1996. Vitamin C, the placebo effect, and the common cold: A case study of how preconceptions influence the analysis of results. *Journal of Clinical Epidemiology* 49:1079–1084.

Hepper, P. G. 1988. The discrimination of human odour by the dog. *Perception* 17: 549–554.

Hilborn, R., and S. C. Stearns. 1982. On inference in ecology and evolutionary biology: The problem of multiple causes. *Acta Biotheoretica* 31:145–164.

Hoffrage, U., S. Lindsey, R. Hertwig, and G. Gigerenzer. 2000. Communicating statistical information. *Science* 290:2261–2262.

Holekamp, K. E., and P. W. Sherman. 1989. Why male ground squirrels disperse. *American Scientist* 77:232–239.

Holmes, D. J., and S. N. Austad. 1995. Birds as animal models for the comparative biology of aging: A prospectus. *Journal of Gerontology* 50A:B59–B66.

Hoover, R. N. 2000. Cancer—nature, nurture, or both. *New England Journal of Medicine* 343:135–136.

Houlahan, J. E., C. S. Findlay, B. R. Schmidt, A. H. Meyer, and S. L. Kuzmin. 2000. Quantitative evidence for global amphibian population declines. *Nature* 404: 752–755.

Hróbjartsson, A., and P. C. Gøtzsche. 2001. Is the placebo powerless? *New England Journal of Medicine* 344:1594–1602.

Hutchins, H. E., and R. M. Lanner. 1982. The central role of Clark's nutcracker in the dispersal and establishment of whitebark pine. *Oecologia* 55:192–201.

IPCC. 2001. *Climate change 2001: The scientific basis. Contribution of Working Group I to the Third Assessment Report of the Intergovernmental Panel on Climate Change*. Cambridge University Press, Cambridge and New York.

Jacobs, L. F., S. J. C. Gaulin, D. F. Sherry, and G. E. Hoffman. 1990. Evolution of spatial cognition: Sex-specific patterns of spatial behavior predict hippocampal size. *Proceedings of the National Academy of Sciences, USA* 87:6349–6352.

Jacobs, L. F., and W. D. Spencer. 1994. Natural space-use patterns and hippocampal size in kangaroo rats. *Brain, Behavior and Evolution* 44:125–132.

Jasanoff, S. 1995. *Science at the bar: Law, science, and technology in America*. Harvard University Press, Cambridge, Mass.

Jee, S. H., J. He, P. K. Whelton, I. Suh, and M. J. Klag. 1999. The effect of chronic coffee drinking on blood pressure: A meta-analysis of controlled clinical trials. *Hypertension* 33:647–652.

Johansen, B. E. 2002. *The global warming desk reference*. Greenwood Press, Westport, Conn.

Johnson, P. T. J., K. B. Lunde, E. G. Ritchie, and A. E. Launer. 1999. The effect of trematode infection on amphibian limb development and survivorship. *Science* 284:802–804.

Johnson, P. T. J., K. B. Lunde, E. M. Thurman, E. G. Ritchie, S. N. Wray, D. R. Sutherland, J. M. Kapfer, T. J. Frest, J. Bowerman, and A. R. Blaustein. 2002.

Parasite (*Ribeiroia ondatrae*) infection linked to amphibian malformations in the western United States. *Ecological Monographs* 72:151–168.

Kaiser, J. 2001. Second look at arsenic finds higher risk. *Science* 293:2189.

Karlowski, T. R., T. C. Chalmers, L. D. Frenkel, A. Z. Kapikian, T. L. Lewis, and J. M. Lynch. 1975. Ascorbic acid for the common cold: A prophylactic and therapeutic trial. *JAMA* 231:1038–1042.

Kemper, A. R., M. M. Davis, and G. L. Freed. 2002. Expected adverse events in a mass smallpox vaccination campaign. *Effctive Clinical Practice* 5:84–90.

Kiesecker, J. M. 2002. Synergism between trematode infection and pesticide exposure: A link to amphibian limb deformities in nature? *Proceedings of the National Academy of Sciences USA* 99:9900–9904.

Kiesecker, J. M., and A. R. Blaustein. 1995. Synergism between UV-B radiation and a pathogen magnifies amphibian embryo mortality in nature. *Proceedings of the National Academy of Sciences USA* 92:11049–11052.

Kiesecker, J. M., A. R. Blaustein, and L. K. Belden. 2001. Complex causes of amphibian population declines. *Nature* 410:681–684.

Klag, M. J., N.-Y. Wang, L. A. Meoni, F. L. Brancati, L. A. Cooper, K.-Y. Liang, J. H. Young, and D. E. Ford. 2002. Coffee intake and risk of hypertension: The Johns Hopkins Precursors Study. *Archives of Internal Medicine* 162:657–662.

Krebs, J. R. 1990. Food–storing birds: Adaptive specialization in brain and behaviour? *Philosophical Transactions of the Royal Society of London Series B—Biological Sciences* 329:153–160.

Krebs, J. R., D. F. Sherry, S. D. Healy, V. H. Perry, and A. L. Vaccarino. 1989. Hippocampal specialization of food-storing birds. *Proceedings of the National Academy of Sciences USA* 86:1388–1392.

Lanner, R. M., and K. F. Connor. 2001. Does bristlecone pine senesce? *Experimental Gerontology* 36:675–685.

Lee, D. W., L. E. Miyasato, and N. S. Clayton. 1998. Neurobiological bases of spatial learning in the natural environment: Neurogenesis and growth in the avian and mammalian hippocampus. *NeuroReport* 9:R15–R27.

LeLorier, J., G. Grégoire, A. Benhaddad, J. Lapierre, and F. Derderian. 1997. Discrepancies between meta-analyses and subsequent large randomized, controlled trials. *New England Journal of Medicine* 337:536–542.

Licht, L. E. 1995. Disappearing amphibians? *Bioscience* 45:307.

———. 1996. Amphibian decline still a puzzle. *Bioscience* 46:172–173.

Lichtenstein, P., N. V. Holm, P. K. Verkasalo, A. Iliadou, J. Kaprio, M. Koskenvuo, E. Pukkala, A. Skytthe, and K. Hemminki. 2000. Environmental and heritable factors in the causation of cancer—analyses of cohorts of twins from Sweden, Denmark, and Finland. *New England Journal of Medicine* 343:78–85.

Light, R. J., and D. B. Pillemer. 1984. *Summing up: The science of reviewing research*. Harvard University Press, Cambridge, Mass.

Linet, M. S. 2000. Evolution of cancer epidemiology. *Epidemiologic Reviews* 22:35–56.

Lorentz, C. P., E. D. Wieben, A. Tefferi, D. A. H. Whiteman, and G. W. Dewald. 2002. Primer on medical genomics. Part I: History of genetics and sequencing of the human genome. *Mayo Clinic Proceedings* 77:773–782.

Macdonald, G. 1961. Epidemiologic models in studies of vector-borne diseases. *Public Health Reports* 76:753–764.

MacDonald, T. M., K. Sharpe, G. Fowler, D. Lyons, S. Freestone, H. G. Lovell, J. Webster, and J. C. Petrie. 1991. Caffeine restriction: Effect on mild hypertension. *British Medical Journal* 303:1235–1238.

Macklin, R. 1999. Ethics, epidemiology, and law: The case of silicone breast implants. *American Journal of Public Health* 89:487–489.

Maguire, E. A., D. G. Gadian, I. S. Johnsrude, C. D. Good, J. Ashburner, R. S. J. Frackowiak, and C. D. Frith. 2000. Navigation-related structural change in the hippocampi of taxi drivers. *Proceedings of the National Academy of Sciences USA* 97: 4398–4403.

Mahlman, J. D. 1998. Science and nonscience concerning human-caused climate warming. *Annual Review of Energy and the Environment* 23:83–105.

Martens, P. 1999. How will climate change affect human health? *American Scientist* 87:534–541.

Martens, P., R. S. Kovats, S. Nijhof, P. de Vries, M. T. J. Livermore, D. J. Bradley, J. Cox, and A. J. McMichael. 1999. Climate change and future populations at risk of malaria. *Global Environmental Change* 9:S89–S107.

Marx, J. 2002. Chromosome end game draws a crowd. *Science* 295:2348–2351.

May, R. M. 1983. Parasitic infections as regulators of animal populations. *American Scientist* 71:36–45.

Mayr, E. 1961. Cause and effect in biology. *Science* 134:1501–1506.

McCauliff, C. M. A. 1982. Burdens of proof: Degrees of belief, quanta of evidence, or constitutional guarantees? *Vanderbilt Law Review* 35:1293–1335.

McGue, M., J. W. Vaupel, N. Holm, and B. Harvald. 1993. Longevity is moderately heritable in a sample of Danish twins born 1870–1880. *Journal of Gerontology* 48:B237–B244.

Medawar, P. B. 1952. An unsolved problem of biology. H. K. Lewis, London.

Medvedev, Z. A. 1990. An attempt at a rational classification of theories of ageing. *Biological Reviews* 65:375–398.

Meydani, M. 2001. Antioxidants and cognitive function. *Nutrition Reviews* 59:S75–S82.

Murphree, A. L. 1997. Retinoblastoma. In D. L. Rimoin, J. M. Connor, and R. E. Pyeritz, eds., *Emery and Rimoin's principles and practice of medical genetics*, pp. 2585–2609. Churchill Livingstone, New York.

National Science Board. 1998. *Science and engineering indicators—1998.* National Science Foundation, Arlington, Virginia.

Newman, B., R. C. Millikan, and M.-C. King. 1997. Genetic epidemiology of breast and ovarian cancers. *Epidemiologic Reviews* 19:69–79.

Nottebohm, F. 1981. A brain for all seasons: Cyclical anatomical changes in song control nuclei of the canary brain. *Science* 214:1368–1370.

O'Brien, S. J., and M. Dean. 1997. In search of AIDS-resistance genes. *Scientific American* 277(3):44–51.

Park, R. L. 2000. *Voodoo science: The road from foolishness to fraud.* Oxford University Press, New York.

Pauling, L. 1970. *Vitamin C and the common cold.* W. H. Freeman, San Francisco.

———. 1971. The significance of the evidence about ascorbic acid and the common cold. *Proceedings of the National Academy of Sciences USA* 68:2678–2681.

Pennisi, E. 2001. The human genome. *Science* 291:1177–1180.

Perls, T. T., L. Alpert, and R. C. Fretts. 1997. Middle-aged mothers live longer. *Nature* 389:133.

Perls, T. T., E. Bubrick, C. G. Wager, J. Vijg, and L. Kruglyak. 1998. Siblings of centenarians live longer. *The Lancet* 351:1560.

Perrig, W. J., P. Perrig, and H. B. Stähelin. 1997. The relation between antioxidants and memory performance in the old and very old. *Journal of the American Geriatrics Society* 45:718–724.

Peto, R. 2000. Cancer, genes, and the environment. *New England Journal of Medicine* 343:1495.

Piedbois, P., and M. Buyse. 2000. Recent meta-analyses in colorectal cancer. *Current Opinion in Oncology* 12:362–367.

Platt, J. R. 1964. Strong inference. *Science* 146:347–353.

Pool, R. 1991. Science literacy: The enemy is us. *Science* 251:266–267.

Popper, K. R. 1965. *Conjectures and refutations: The growth of scientific knowledge.* Basic Books, New York.

Pravosudov, V. V., and N. S. Clayton. 2002. A test of the adaptive specialization hypothesis: Population differences in caching, memory, and the hippocampus in black-capped chickadees (*Poecile atricapilla*). *Behavioral Neuroscience* 116:515–522.

Puca, A. A., M. J. Daly, S. J. Brewster, T. C. Matise, J. Barrett, M. Shea-Drinkwater, S. Kang, E. Joyce, J. Nicoli, E. Benson, L. M. Kunkel, and T. Perls. 2001. A genome-wide scan for linkage to human exceptional longevity identifies a locus on chromosome 4. *Proceedings of the National Academy of Sciences USA* 98:10505–10508.

Reboreda, J. C., N. S. Clayton, and A. Kacelnik. 1996. Species and sex differences in hippocampus size in parasitic and non-parasitic cowbirds. *Neuroreport* 7:505–508.

Reeburgh, W. S. 1997. *Figures summarizing the global cycles of biologically active elements.* http://ess1.ps.uci.edu/~reeburgh/figures.html (accessed April 3, 2003).

Rensberger, B. 2000. The nature of evidence. *Science* 289:61.

Rice, W. R. 1989. Analyzing tables of statistical tests. *Evolution* 43:223–225.

Ricklefs, R. E. 1998. Evolutionary theories of aging: Confirmation of a fundamental prediction, with implications for the genetic basis and evolution of life span. *The American Naturalist* 152:24–44.

Ricklefs, R. E., and C. E. Finch. 1995. *Aging: A natural history.* W. H. Freeman, New York.

Risch, N. 2001. The genetic epidemiology of cancer: Interpreting family and twin studies and their implications for molecular genetic approaches. *Cancer Epidemiology, Biomarkers & Prevention* 10:733–741.

Robinson, D. 2002. *Cancer clusters: Findings vs feelings.* Medscape General Medicine 4(4). http://www.medscape.com/viewarticle/442554 (accessed November 23, 2002).

Roeder, K. 1994. DNA fingerprinting: A review of the controversy. *Statistical Science* 9:222–247.

Rogers, D. J., and S. E. Randolph. 2000. The global spread of malaria in a future, warmer world. *Science* 289:1763–1766.

Rose, G. 1981. Strategies of prevention: Lessons from cardiovascular disease. *British Medical Journal* 282:1847–1851.

———. 1992. *The strategy of preventive medicine.* Oxford University Press, Oxford.

Rose, M. R. 1984. Laboratory evolution of postponed senescence in *Drosophila melanogaster. Evolution* 38:1004–1010.

Rosmarin, P. C., W. B. Applegate, and G. W. Somes. 1990. Coffee consumption and blood pressure: A randomized, crossover clinical trial. *Journal of General Internal Medicine* 5:211–213.

Royall, R. M. 1991. Ethics and statistics in randomized clinical trials (with discussion). *Statistical Science* 6:52–88.

Sano, M., C. Ernesto, R. G. Thomas, M. R. Klauber, K. Schafer, M. Grundman, P. Woodbury, J. Growdon, C. W. Cotman, E. Pfeiffer, L. S. Schneider, and L. J. Thal. 1997. A controlled trial of selegiline, alpha-tocopherol, or both as treatment for Alzheimer's disease. *New England Journal of Medicine* 336:1216–1222.

Sarkar, S. 1996. Ecological theory and anuran declines. *Bioscience* 46:199–207.

Schoon, G. A. A. 1996. Scent identification lineups by dogs (*Canis familiaris*): Experimental design and forensic application. *Applied Animal Behaviour Science* 49: 257–267.

———. 1997. Scent identification by dogs (*Canis familiaris*): A new experimental design. *Behaviour* 134:531–550.

———. 1998. A first assessment of the reliability of an improved scent identification line-up. *Journal of Forensic Science* 43:70–75.

Senior, K. 2002. FDA panel rejects common cold treatment. *The Lancet Infectious Diseases* 2:264.

Sessions, S. K., and S. B. Ruth. 1990. Explanation for naturally occurring supernumerary limbs in amphibians. *Journal of Experimental Zoology* 254:38–47.

Settle, R. H., B. A. Sommerville, J. McCormick, and D. M. Broom. 1994. Human scent matching using specially trained dogs. *Animal Behaviour* 48:1443–1448.

Shanley, D. P., and T. B. L. Kirkwood. 2001. Evolution of the human menopause. *BioEssays* 23:282–287.

Sharpe, D. 1997. Of apples and oranges, file drawers and garbage: Why validity issues in meta-analysis will not go away. *Clinical Psychology Review* 17:881–901.

Sherman, P. W. 1988. The levels of analysis. *Animal Behaviour* 36:616–619.

———. 1998. The evolution of menopause. *Nature* 392:759–760.

Sherman, P. W., and S. M. Flaxman. 2001. Protecting ourselves from food. *American Scientist* 89:142–151.

———. 2002. Nausea and vomiting of pregnancy in an evolutionary perspective. *American Journal of Obstetrics and Gynecology* 186:S190–S197.

Sherry, D. F., A. L. Vaccarino, K. Buckenham, and R. S. Herz. 1989. The hippocampal complex of food-storing birds. *Brain, Behavior and Evolution* 34:308–317.

Shettleworth, S. J. 1990. Spatial memory in food-storing birds. *Philosophical Transactions of the Royal Society of London Series B—Biological Sciences* 329:143–151.

Sinclair, A. J., A. J. Bayer, J. Johnston, C. Warner, and S. R. Maxwell. 1998. Altered plasma antioxidant status in subjects with Alzheimer's disease and vascular dementia. *International Journal of Geriatric Psychiatry* 13:840–845.

Souder, W. 2000. *A plague of frogs: The horrifying true story.* Hyperion, New York.

Steel, C. M. 1997. Cancer of the breast and female reproductive tract. In D. L. Rimoin, J. M. Connor, and R. E. Pyeritz, eds., *Emery and Rimoin's principles and practice of medical genetics*, 3rd ed., pp. 1501–1523. Churchill Livingstone, New York.

Stein, Z. A. 1999. Silicone breast implants: Epidemiological evidence of sequelae. *American Journal of Public Health* 89:484–487.

Stevens, T. A., and J. R. Krebs. 1986. Retrieval of stored seeds by marsh tits *Parus palustris* in the field. *Ibis* 128:513–525.

Superko, H. R., W. Bortz, Jr., P. T. Williams, J. J. Albers, and P. D. Wood. 1991. Caffeinated and decaffeinated coffee effects on plasma lipoprotein cholesterol, apolipoproteins, and lipase activity: A controlled, randomized trial. *American Journal of Clinical Nutrition* 54:599–605.

Superko, H. R., J. Myll, C. DiRicco, P. T. Williams, W. M. Bortz, and P. D. Wood. 1994. Effects of cessation of caffeinated-coffee consumption on ambulatory and resting blood pressure in men. *American Journal of Cardiology* 73:780–784.

Sutera, R. 2000. The origin of whales and the power of independent evidence. *Reports of the National Center for Science Education* 20(5):33–41.

Swift, M., D. Morrell, R. B. Massey, and C. L. Chase. 1991. Incidence of cancer in 161 families affected by ataxia telangiectasia. *New England Journal of Medicine* 325:1831–1836.

Taslitz, A. E. 1990. Does the cold nose know? The unscientific myth of the dog scent lineup. *Hastings Law Journal* 42:15–134.

Taubes, G. 1998. The (political) science of salt. *Science* 281:898–907.

Tavani, A., and C. La Vecchia. 2000. Coffee and cancer: A review of epidemiological studies, 1990–1999. *European Journal of Cancer Prevention* 9:241–256.

Taylor, K. 1999. Rapid climate change. *American Scientist* 87:320–327.

Terry, P., L. Bergkvist, L. Holmberg, and A. Wolk. 2001. Coffee consumption and risk of colorectal cancer in a population based prospective cohort of Swedish women. *Gut* 49:87–90.

Thewissen, J. G. M. 1998. *The emergence of whales: Evolutionary patterns in the origin of Cetacea.* Plenum, New York.

Tinbergen, N. 1963. On aims and methods of ethology. *Zeitschrift fur Tierpsychologie* 20:410–433.

Trichopoulos, D., F. P. Li, and D. J. Hunter. 1996. What causes cancer? *Scientific American* 275(3):80–87.

Tudge, C. 2002. Why science should warm our hearts. *Reports of the National Center for Science Education* 22(1–2):41–44.

Tyner, S. D., S. Venkatachalam, J. Choi, S. Jones, N. Ghebranious, H. Igelmann, X. Lu, G. Soron, B. Cooper, C. Brayton, S. H. Park, T. Thompson, G. Karsenty, A. Bradley, and L. A. Donehower. 2002. p53 mutant mice that display early ageing-associated phenotypes. *Nature* 415:45–53.

United States Census Bureau. 2002. Vital statistics No. 98. Expectation of life and expected deaths by race, sex, and age: 1998. In 2001 *Statistical Abstract of the United States, Section 2*, p. 74. http://www.census.gov/prod/2002pubs/01statab/vitstat.pdf (accessed April 3, 2003).

van Dusseldorp, M., P. Smits, J. W. M. Lenders, L. Temme, T. Thien, and M. B. Katan. 1992. Effects of coffee on cardiovascular responses to stress: A 14-week controlled trial. *Psychosomatic Medicine* 54:344–353.

van Dusseldorp, M., P. Smits, T. Thien, and M. B. Katan. 1989. Effect of decaffeinated coffee versus regular coffee on blood pressure: A 12-week, double-blind trial. *Hypertension* 14:563–569.

Vander Wall, S. B. 1982. An experimental analysis of cache recovery in Clark's nutcracker. *Animal Behaviour* 30:84–94.

———. 1990. *Food hoarding in animals.* University of Chicago Press, Chicago.

———. 1998. Foraging success of granivorous rodents: Effects of variation in seed and soil water on olfaction. *Ecology* 79:233–241.

———. 2000. The influence of environmental conditions on cache recovery and cache pilferage by yellow pine chipmunks (*Tamias amoenus*) and deer mice (*Peromyscus maniculatus*). *Behavioral Ecology* 11:544–549.

Vander Wall, S. B., and R. P. Balda. 1977. Coadaptations of the Clark's nutcracker and the piñon pine for efficient seed harvest and dispersal. *Ecological Monographs* 47:89–111.

Varmus, H., and R. A. Weinberg. 1993. *Genes and the biology of cancer.* Freeman, New York.

Vogel, F., and A. G. Motulsky. 1997. *Human genetics: Problems and approaches*, 3rd ed. Springer-Verlag, Berlin.

Ware, J. H. 1989. Investigating therapies of potentially great benefit: ECMO (with discussion). *Statistical Science* 4:298–340.

Weinberg, R. A. 1996. How cancer arises. *Scientific American* 275(3):62–71.

Wells, D. L., and P. G. Hepper. 2000. The discrimination of dog odours by humans. *Perception* 29:111–115.

Wells, G. L., R. S. Malpass, R. C. L. Lindsay, R. P. Fisher, J. W. Turtle, and S. M. Fulero. 2000. From the lab to the police station—a successful application of eyewitness research. *American Psychologist* 55:581–598.

Willett, W. C. 2002. Balancing life-style and genomics research for disease prevention. *Science* 296:695–698.

Wilson, E. O. 2002. *The future of life.* Knopf, New York.

Wong, K. 2002. The mammals that conquered the seas. *Scientific American* 286(5): 70–79.

Yang, Q., and M. J. Khoury. 1997. Evolving methods in genetic epidemiology. III. Gene-environment interaction in epidemiologic research. *Epidemiologic Reviews* 19:33–43.

Index

Page numbers in italics refer to figures and tables.